MRCPsych CASC Notes

Arpan Dutta

Nishan M. Bhandary

To those who believed in us and helped us succeed.

PREFACE

The Clinical Asessment of Skills and Competencies (CASC) exam of the Royal College of Psychiatrists was designed to assess the clinical competencies expected of trainees after 30 months of training. There are 2 "circuits" as the Royal College describes them. The first circuit has 8 stand alone stations of 7 minutes with 1 minute preparation time. The second circuit has 4 pairs of linked stations which last 10 minutes with 2 minutes of 'preparation' time. Each station has a "construct" which sets the scene and tells you the task or tasks to be completed. Although the stations are linked the marks for each station are independent. Currently 12 out of 16 stations need to be passed to pass the CASC overall. Each station is graded as Fail, Borderline Fail, Borderline Pass or Pass. A Borderline Pass and a Pass are both considered a Pass. Common "Areas of Concern" are listed on the Royal College of Psychiatrists Website (www.rcpsych.ac.uk/exams/about/mrcpsychcasc.aspx)

This book is the result of our studies which we began in January 2009. It is the compilation of work from a number of sources as well as our own knowlegde. The intention of this book is to help those pass what can appear a difficult exam and provide the necessary theory. It is impossible to cover all the mental disorders and speculate on the possible case vignettes. Hence we have concentrated on what we believe are the likely stations that could come up in the exam.

SUGGESTIONS FOR PRACTICE

Medicine and psychiatry are practical subjects and the exam can only be passed through arduous practice. The CASC exam calls for adaptability and a conversational style that allows for better collection of information, this cannot be achieved by the use of "stock phrases". The Royal College suggests that it should be "focussed, fluent, and demonstrates empathy with the patient's experience. They should display an appropriate mix of open and closed questioning and display advanced listening skills. A "checklist" approach to history taking should not be rewarded".

We suggest practicing in pairs. More than 2 people can lead to problems with managing the time taken to complete the variety of possible permutations and combinations of possible stations

For each station that you encounter there are a few things that must be done. Again this is from our own experience:

> Finish all of the tasks asked- if you do not complete the tasks you cannot be passed

- Biological, Psychological and Social elements in all management plans and in history taking

- Think about who you are talking to- consultant/family/patient and vary language accordingly

- Maximum of 3 questions in the last minute- this will vary by station and sitiuation

- Crystallise the diagnosis as much as possible

- Whilst talking to nurses and managers, talk to them as part of a multidisciplinary team, and treat them as equals

- In all physical examination stations you should use the alcohol gel to cleanse your hands. Not using the alcohol gel will not cause you to fail accord to the Royal College but will mean that the examiner will mark you one grade lower than you would have been marked.

Good luck!

Arpan Dutta & Nishan M Bhandary

ABOUT THE AUTHORS

Arpan Dutta completed his undergraduate medical training from the University of Liverpool, England in 2004 passing with honours. He did his core psychiatric training in Merseyside. He passed the MRCPsych in May 2009. He is currently working as Speciality Registrar Year 4 in General Adult Psychiatry in Cheshire.

Nishan M Bhandary completed his undergraduate training at the Kasturba Medical College, Manipal Academy of Higher Education University, Mangalore, India. He finished his core psychiatric training in Mersey Deanery, Liverpool and passed his MRCPsych in October 2009. He is currently working as Speciality Registrar Year 4 in General Adult Psychiatry in Liverpool.

CONTENTS

Chapter 1   Psychopathology   15

Chapter 2   Suicide/DSH- risk assessment   23

Chapter 3   Organic Conditions   27

Chapter 4   Alcohol and Drugs   63

Chapter 5   Schizophrenia & Psychosis   71

Chapter 6   Mood Disorders   87

Chapter 7   Anxiety Disorders   117

Chapter 8   Eating Disorders   149

Chapter 9   Puerperal Disorders   161

Chapter 10   Personality Disorder   173

Chapter 11   Forensic Psychiatry   185

Chapter 12   Learning Disability   193

Chapter 13   Childhood Disorders   195

Chapter 14   Physical Examinations   217

Chapter 15   Psychotherapy   233

Chapter 16   Miscellaneous   243

Chapter 17   Incidence, prevalence, prognosis   251

References   265

Index   270

CHAPTER 1   PSYCHOPATHOLOGY

This chapter deals with the questions in eliciting basic psychopathology which is the crucial element in any history taking station. The aim is not to ask all the questions listed below, but asking an open question initially and then focussing down to the diagnosis accordingly. We have concentrated mainly on delusions, hallucinations and insight with the rest covered in the relevant chapters.

Remember: Do not take a checklist approach to asking questions. Try to relate to patient's experience.

DELUSIONS

Ideas of reference:

Do you think people talk behind your back?

Do you ever get the feeling that other people are taking notice of you, example, when you are walking in the street?

Do you see any reference to yourself on TV or in newspapers?

Do they ever seem to laugh at you or talk about you critically?

Delusional mood:

Do you ever get the feeling something unusual is happening which you cannot explain?

Delusional perception:

Do you think things happening around you have a special meaning to you?

Has a sudden explanation occurred to you out of the blue?

Delusion:

Do you have beliefs that other people have considered odd or unusual?

- ➢ Need to know its fixed
- ➢ Need to know its false
- ➢ Need to know that others don't believe the same
- ➢ Try to differentiate it from overvalued ideas and assess the degree of conviction- Could there be any other explanation? Is there any possibility that you could be wrong about this?

Delusions of control:

Do you think you are in control of yourself?

Do you feel under the control of some force or power other than yourself?

Delusions of persecution:

Is anyone deliberately trying to harm you? (example- trying to poison you or kill you)

Grandiose delusions:

Do you think there is anything special about you?

Do you think you have special abilities or powers?

Do you think you are a very prominent person or related to someone prominent, like Royalty?

Is there a special purpose or mission to your life?

Religious delusions:

Are you a very religious person?

Are you specially close to god?

Can god commmunicate with you? If yes, how?

Delusions of guilt:

Do you feel that you have committed a crime, or sinned greatly, or deserve punishment?

Nihilistic delusions:

Do you think something terrible has happened or will happen to you?

Delusional jealousy:

Do you have any reason to be jealous of anybody? Are you worried about your partner being faithful?

Hypochondriachal delusions:

Do you think you have a serious illness?

HALLUCINATIONS

When asking about hallucinations you need to know:

- Onset
- Duration
- Is it in Internal/external space (inside/outside your head)
- Does it feel real or not
- $2^{nd}/3^{rd}$ person
- Content
- Open to conscious manipulation? (can you control it?)
- Modalities
  - Functional
  - Reflex

- Extracampine

- Visual, olfactory, gustatory, tactile

➢ Command hallucinations

Start by gently leading into your question.

"I would like to ask you a routine question which we ask of everybody. Have you noticed anything unusual recently? For example, do you ever seem to hear noises or voices when there is no one about, and nothing else to explain it?"

Auditory hallucinations:

Does it sound like muttering or whispering? (Is it as clear as I am talking to you?)

Do you hear several voices?

Do they speak directly to you?

What do they say? (Which language?)

Do they speak among themselves? (third person)

Do these voices describe or comment upon what you are doing or thinking? (running commentary)

Do the voices tell you to do things (or give you orders)? (command hallucinations)

Do you feel compelled to obey them?

If accusatory- Do you think that it is justified? Do you deserve it?

Visual hallucinations:

Have you had visions or seen things that other people couldn't see? (if yes, explore further- when, where, ask for explanation)

Olfactory hallucinations:

Do you sometimes notice strange smells that other people don't notice?

Tactile hallucinations:

Do you ever feel that someone is touching you, but when you look there is nobody there? (for example sensations like insects crawling or electricity passing?)

FIRST RANK SYMPTOMS

Third person auditory hallucinations

Running commentary

Thought echo: Do you ever seem to hear your own thoughts repeated or echoed?

Thought alienation (general question)- Are you able to think clearly? Are you in full control of your thoughts?

Thought insertion: Are thoughts put into your head which you know are not your own?

Thought withdrawal: Do your thoughts ever seem to be taken out of your head as though some external person or force were removing them?

Thought broadcast: Do you think your thoughts can be read by others?

Thought block: Do you ever get the experience of your thoughts stopping quite unexpectedly?

Made will (volition): Do you think you are being controlled by someone/ something? Does this person/force sometimes force its feelings on you against your will?

Made acts: Do you feel that your actions are being executed by an external influence?

Made affect: Is there someone/something trying to change the way you feel in yourself?

Somatic passivity: Have you felt strange experiences in your body?

INSIGHT

What do you think is happening?

Could these symptoms be due to mental illness?

Do you think medication/ treatment might help?

Are you willing to accept help?

CHAPTER 2   SUICIDE/DSH- RISK ASSESSMENT

Features of the act:

- Method (tablets- how many, dose, where did you get the tablets from) with alcohol/drugs?

- Triggers- Clear precipitant, intoxicated, direct gain?

- Patient's belief in the lethality- Did you think it would kill you?

- Length of planning- Impulsive or planned out

- Final acts- Suicide note/text/telephone message/any other acts of closure (eg-thinking of children, will, etc.)

- Precautions to avoid discovery- Where did it take place/Did you anticipate being found/Did you inform anyone/Was anyone else present?

- Previous similar acts (different features?)

– Actions after the act- What happened? How did you end up in hospital?

Mental state:

- Wish to die?

- Cry for help? (Was it the only way you could tell how you felt?)

- Attitude to survival- Relieved or disappointed to be alive?/Ongoing wish to die?/ future plans

- Affective symptoms (symptoms associated with suicide- hopelessness, suicidal ideation, depression, anhedonia, insomnia, severe anxiety, impaired concentration, psychomotor agitation, panic attacks)

- Substance misuse problems (drugs and alcohol)

- Other mental disorder

- Risk to others- Does this act put anyone else at risk?

Past psychiatric history

Medical history

Family history of mental illness and suicide

Personal history including relationships and work

Social history including recent life events and support system

Forensic history

Premorbid personality- coping with problems?/personality disorder

INDICATORS OF SERIOUSNESS

- Act involved a dangerous method

- Detailed planning

- High subjective impression of lethality

- Precautions to avoid discovery

- Didn't seek help after the act

- Final acts

- Regret on realizing they are alive

In the case of overdose following sexual assault or rape, you will need to know who did it for the sake of management plan....i.e. is there anyone she is scared of at home? (and therefore does appropriate support/accommodation need to be arranged?) Consider risk to others and protective factors. You will also need to address confidentiality.

SUICIDE- HIGH RISK FACTORS

- ➤ Male
- ➤ Elderly- but also high rates in young males
- ➤ Living alone
- ➤ Presence of stressful life events
- ➤ Psychiatric or physical disorder
- ➤ Substance misuse
- ➤ Borderline personality disorder
- ➤ History of suicide attempts
- ➤ History of DSH
- ➤ Family history of suicide

CHAPTER 3   ORGANIC MENTAL DISORDERS

ALZHEIMER'S DEMENTIA

a) Dementia (evidence of decline in both memory and thinking which is sufficient to impair personal activities of daily living, for atleast six months)

b) Insidious onset, slow deterioration

c) No organic cause

d) Absence of sudden onset or of neurological signs of focal damage (eg hemiparesis)

VASCULAR DEMENTIA

a) Dementia

b) Uneven cognitive impairment

c) Abrupt onset or a stepwise deterioration

d) Focal neurological signs and symptoms

e) Associated features are hypertension, carotid bruit, emotional lability with transient depressive mood, weeping/explosive laughter, and transient episodes of clouded consciousness/delirium often provoked by further infarction. Personality is relatively well preserved although may present with apathy, disinhibition, or accentuation of previous traits such as egocentricity, paranoid attitudes, or irritability.

DEMENTIA IN PICK'S DISEASE

a) Progressive dementia

b) Predominant frontal lobe features with euphoria, emotional blunting, and coarsening of social behaviour, disinhibition and apathy or restlessness

c) Behavioural manifestations which commonly precede frank memory impairment

DEMENTIA IN HUNTINGTON'S DISEASE

a) Symptoms typically emerge in the 3rd or 4th decade and progression is slow.

b) Association of choreiform movement disorder, dementia, and family history of Huntington's disease is highly suggestive.

c) It is characterised by the predominant involvement of frontal lobe functions in the early stage, with relative preservation of memory until later.

ORGANIC AMNESIC SYNDROME (INCLUDING KORSAKOFF'S SYNDROME)

a) Presence of a memory impairment manifest in a defect of recent memory (impaired learning of new material); anterograde and retrograde amnesia and a reduced ability to recall past experiences in reverse order of their occurence.

b) History of objective evidence of an insult to, or a disease of, the brain

c) Absence of a defect in immediate recall, of disturbances of attention and consciousness and of global intellectual impairment.

DELIRIUM

a) Impairment of consciousness and attention

b) Global disturbance of cognition (perceptual distortions, illusions and hallucinations most often visual; impairment of immediate recall, disorientation to time, place or person)

c) Psychomotor disturbances (hypo/hyperactivity; increased reaction time, increased or decreased flow of speech; enhanced startle reaction) disturbance of the sleep wake cycle (insomnia, total sleep loss or reversal); day time drowsiness; nocturnal worsening of symptoms; disturbing dreams or nightmares, which may continue as hallucinations after awakening

d) Emotional disturbances (eg depression, anxiety)

The onset is is usually rapid, the course diurnally fluctuating, and the total duration lasts less than 6 months.

MILD COGNITIVE DISORDER

The main feature is a decline in cognitive performance. This may include memory impairment, learning or concentration difficulties. Objective tests usually indicate abnormality. The symptoms are such that a diagnosis of dementia, organic amnesic syndrome or delirium cannot be made.

ORGANIC PERSONALITY DISORDER (INCLUDES FRONTAL LOBE SYNDROME)

Established history or other evidence of brain disease, damage, or dysfunction, a definitive diagnosis requires the presence of two or more of the following features:

a) Reduced ability to persevere with goal directed activities

b) Altered emotional behaviour, characterised by emotional lability, shallow and unwarranted cheerfulness, and easy change to irritability or short lived outbursts of anger and aggression, apathy

c) Expression of needs and impulses without consideration of consequences of social convention (dissocial acts such as stealing, inappropriate sexual advances, or voracious eating, disregard for personal hygiene)

d) Cognitive disturbances (suspiciousness, paranoid ideation)

e) Marked alteration of the rate and flow of language production, with features such as circumstantiality, overinclusiveness

f) Altered sexual behaviour

LEWY BODY DEMENTIA

a) Progressive cognitive decline

b) 2 or more of fluctuating cognition with pronounced variation in cognition/alertness, recurrent complex visual hallucinations, spontaneous motor features of parkinsonism

c) Supportive features include repeated falls, syncope, neuroleptic sensitivity, systematised delusions, hallucinations in other modalities

DEMENTIA- HISTORY TAKING

(This is usually collateral history from the carer)

Onset and progression- sudden or gradual?

What and when was it first noticed?

Examples for memory loss?

Fluctuations?

Cognitive symptoms:

a) memory (short and long term)
   - Can he/she remember things that happened in the last few minutes/ in the day/few years ago?
   - Does prompting help?
   - Is it consistent or patchy?

b) Temporal orientation- Time of day, date, day

c) Spatial orientation- Does he lose his way?

d) Language difficulties- Any changes to the way he speaks, like word finding difficulties?

e) Comprehension- Can he understand when others speak to him?

f) Dyspraxia

- Does he have difficulty doing things?

- Does he have the ability to look after himself?

- Activities of Daily Living (ADL)- personal hygiene, washing, cooking

g) Dyslexia, dysgraphia

h) Visuospatial difficulties, agnosia

- Does he have difficulty in recognising things or places?

- Does he have difficulty in recognising familiar faces?

i) Judgement and decision making

Behavioural symptoms:

a) Aggression

b) Violent outbursts

c) Inappropriate behaviour like disinhibition

d) Socially withdrawn

e) Wandering at night

Psychological symptoms:

a) Depression

b) Anxiety

c) Apathy

d) Psychosis

Physical symptoms:

a) Sensory impairment

b) Weakness of limbs

c) Gait disturbance

d) Abnormal movements (Parkinson's)

e) Incontinence

Risk assessment:

a) Fire risk – using cooker, smoking

b) Management of finances

c) Inappropriate use/forgetting of medications

d) Letting strangers inside the house

e) Risk factors for dementia like alcohol and head injury

f) Driving

Other relevant factors:

a) Current medications

b) Past psychiatric history

c) Past medical history like diabetes, hypertension, thyroid problems, stroke and infections

d) Family history

e) Personal history including occupation and social history

MANAGEMENT OF DEMENTIA- DISCUSS WITH CONSULTANT

(Modify language if discussing with carer)

a) Summarise the history, mental state examination and risk assessment

b) Mention the differential diagnosis and the most likely working diagnosis

c) Would like to gather collateral history from family and the GP

d) Routine blood tests as part of dementia screening

e) Screen for vascular risk factors

f) Consider EEG for delirium, FTD, CJD

g) Consider CT/MRI/SPECT scan to reveal vascular lesions (see NICE guidelines)

h) Neuropsychological assessment including MMSE (beware of the pitfalls in MMSE)

i) Follow up in the community or admission to old age ward for further assessment

j) Management includes biological, psychological and social approach in the immediate and longer term

- Review MMSE, global, functional and behavioural assessment every 6 months
- CPN visits to patient
- Referral to social services – help with carers; day centres – so carer can get a break; meals on wheels; referral to OT – bath equipment, banister, support rails, wheelchair; short stay of 1-2 weeks in residential home to provide relatives a short period of respite; may need management in a suitable residential or nursing home at some stage
- Financial and legal arrangements(e.g.- use of advocacy)
- More information - Alzheimers society, Age Concern, Carers UK

SELF HELP

- Take notice – (e.g. repeating the name of the person you just met)
- Take notes
- Diary
- Get organized (more likely to remember things if you are organized)
- Keep fit
- Regular health checks
- Reality orientation (helping to remember where things are, day date, time, what is happening)
- Mnemonics
- External aids – alarms, calendars

FOR CARERS

- Keep things in familiar places.

- Have clocks/calendars showing correct date and time

- Message board for important dates and appointments

- Make a list of daily activities

- Establish a route and leave a note if you are going out

- Making an album of photos of families and friends labelled with names

- Make sure that they can see, hear and speak as well as possible (check hearing aids, glasses and dentures).

- Attract attention before you speak, perhaps by gently touching their arm.

- Try to avoid other distractions such as from a television

- While you are conversing try to keep your head and shoulders at the same level.

- Try to maintain eye contact when either of you is speaking (this will help keep her attention).

- Hold their hand during your conversation.

- Try to remain as calm as possible.

- Speak as clearly as you can.

- Use short sentences and try to talk about only one thing at a time.

- Write things down if it helps.

ANTIDEMENTIA DRUGS

Mechanism of action:

The main drugs licenced for Alzheimers dementia are acetylcholinesterase inhibitors. In Alzheimers dementia, a chemical in the brain called acetylcholine, which is important for learning and memory, is in short supply. These drugs increase the levels of acetylcholine.

Drug names:

Donepezil (Aricept)- 5mg OD, increased to max dose 10mg OD (as necessary)

Galantamine (Reminyl)- 4mg BD, increased to max dose 12mg BD (as necessary)

Rivastigmine (Exelon)- 1.5mg BD, increased to max dose 6mg BD (as necessary)

Effect:

They improve memory particularly remembering new info as well as recalling old info. They also have general benefits including improving alertness and motivation. It may take some months of treatment for there to be a noticeable improvement or slowing down of memory loss. Some report improved mood and will be able to perform tasks which they had forgotten such as shopping. 5-6 out of 10 people will show improvement or stabilization.

Side effects:

Common- nausea, tiredness, loss of appetite, diarrhea, muscle cramps, poor sleep. It is recommended to take with food and the side effects will fade after few weeks and will go away if the medicine is stopped.

Rarely can cause urinary retention and seizures.

Contraindications (CI):

Relative CI include asthma, peptic ulcers, COPD, supraventricular conduction abnormalities

Who can prescribe:

Started by specialist rather than GP. You will be monitored in the memory clinic and after starting the medication you will be reviewed at 4 weeks. CPN will review at 2 weeks to make sure no side effects.

How long to be taken?

Unfortunately not everyone benefits and it is advised that they should be used only in the early stages of Alzheimers' dementia and should only be given in

38

mild/moderate stages only. We will continue the treatment if the MMSE score remains at or above 10 points and if there is a noticeable improvement in the function and behaviour of the patient.

Interactions:

No problems with alcohol. Can cause drowsiness but not a main side effect.

No dietary restrictions.

No problems with other medications

Memantine (Ebixa)

It is thought to work by affecting glutamate, a brain chemical which is involved in learning and memory. In Alzheimers dementia too much glutamate leaks out of damaged brain cells and this interferes with learning and memory. In studies completed so far just over half the people taking memantine show some slowing down in the progression of dementia but this effect has only been demonstrated in more severe dementia.

Side effects- nausea, restlessness, stomach ache, headache

Ginko Biloba

It is a naturally occurring substance extracted from the maidenhair tree. Some studies show that they are nearly as effective as cholinesterase inhibitors. They are thought to work as an antioxidant (clears free radicals released by damaged cells), also improves circulation. Mainly used in combination with other treatment

Side effects- delay blood clotting.

We would advise that you first discuss this with the doctor You can get it without a prescription. Ensure standardized extract of the leaf rather than leaf powder therefore standardized dose.

Vitamin E

It is a natural substance found in oil from soya bean, sunflower seeds, corn seed whole grain foods, fish oil, nuts. It acts as an antioxidants. Some studies demonstrate it slows the progression of AD. It can interfere with blood clotting. Dose should not exceed more than 200 U/day.

Selegiline

This is a drug normally used in Parkinsons disease. It acts as an antioxidant. Side effects include low BP, nausea, dizziness and vivid dreams.

OTHER THERAPIES

- Aromatherapy (use of fragrant oils to relax)
- Multisensory stimulation (snoezelan room – special room designed to gently stimulate the senses while also helping agitated people to relax. There are comfortable places to sit, coloured restful lights and restful music)
- Therapeutic use of music and dance
- Animal assisted therapy
- Massage
- Reminiscence therapy – stimulating the recollection of events or memories from the patient achieved by music, videotapes or pictures

ANTIPSYCHOTICS IN DEMENTIA

Don't use antipsychotics for mild to moderate symptoms in DLB, alzheimers, vascular or mixed dementias – but it is considered for psychosis/agitation.

Risks and benefits balance should be fully discussed. There are conerns with its use but we have to weigh up risks and benefits. Referral to other options may take take time and not work as quickly as the medication.

It is given for a short period of time, low dose for agitation.

Incidence of stroke is 1-2%, 3-4 times compared to others who do not have dementia.

LEWY BODY DEMENTIA MANAGEMENT

There can be severe sensitivity reactions (in up to 50% of those treated with neuroleptics) including precipitation of irreversible parkinsonism and autonomic disturbances similar to NMS

Antipsychotics should be used with caution and commenced at low dose. Currently only quetiapine and clozapine are recommended.

Rivastigmine is licensed for DLB. It helps with cognition and with delusions and hallucinations. It is well tolerated but minority develop tremor or psychomotor agitation.

Dopamine agonists or anticholinergic medication to treat parkinsonian symptoms tends to considerably worsen psychotic symptoms in lewy body dementia

EXPLAIN ALZHEIMER'S DISEASE

Dementia is a progressive, irreversible disease which affects the working functions of the brain causing confusion, changes in personality and behaviour and memory. Memory is usually affected first.

Whats is Alzheimer's dementia?

It is the commonest type of dementia (70% of dementia cases). Everyone loses brain cells as they get older. In Alzheimer's, this process is more severe and rapid than ageing. The onset tends to be gradual.

How common?

5 out of 100 people over the age of 65 get affected. Both males and females are equally affected. Sometimes, it can also occur at a younger age.

Causes:

- It tends to run in families.
- It is very common in sufferers with Down's syndrome.
- Hypertension, diabetes, increased cholesterol, smoking, alcohol and being over weight- increase the risk by causing problems with the blood supply to the brain.
- A severe head injury at some point in life may increase the risk.

Symptoms:

As mentioned in history taking

Progress:

Dementia nearly always gets steadily worse. It may take a few years, may progress quickly but more often gradual. Unfortunately it is a progressive condition. It cannot be halted or reversed. Ultimately I am afraid it will deteriorate. Most studies show people live 5-10 years after being diagnosed. What we can assure you is that we will do everything possible to keep him healthy.

Management:

As mentioned earlier

EXPLAIN VASCULAR DEMENTIA

Arteries supplying blood to the brain become blocked., leading to small strokes. Parts of brain die as they are starved of oxygen. This is more likely if you have high BP, diabetes, high cholesterol. A stroke can be so slight that it causes no immediate symptoms, or may just cause a brief spell of dizziness, weakness or confusion. Eventually this damage accumulates sufficiently to cause dementia.

Accounts for 1 in 5 cases of dementia

EXPLAIN LEWY BODY DEMENTIA

People with Lewy body dementia have symptoms which overlap with Alzheimer's disease and Parkinson's disease. The level of confusion can vary during the course of the day, but visual hallucinations of people or animals are more common. They may also have a tremor, muscle stiffness, falls or difficulty with walking.

EXPLAIN FRONTO-TEMPORAL DEMENTIA

If the dementia affects the front of the brain more than other areas, it is more likely to cause personality changes as well as memory problems. Usually progressive personality change without prominent memory problems early in the course of the illness – table manners deteriorate, preference for sweet foods, hoarding, disihibition.

BEHAVIOURAL SYMPTOMS IN DEMENTIA

Elicit history in a temporal sequence, concentrating on **A**ntecedent, **B**ehaviour and **C**onsequence.

Biological causes –infection, constipation, sensory impairment, weakness, gait disturbance, abnormal movements, incontinence, medication changes, pain

Psychological causes– behavioural (aggression, violent outbursts, inappropriate behaviour), depression, anxiety, psychosis, cognitive symptoms (dyspraxia, dyslexia)

Social causes – loss of mobility, loss of access to activities, cessation of contact with carers, change of staff or residents, physical environment – overcrowding, lack of privacy, inadequate staff attention, poor communication

DEMENTIA AND WANDERING – ASSESS RISKS

Current episode of wandering:

- Where was the patient found?
- What was the patient trying to do?
- Was the patient appropriately dressed – shoes, coat etc?
- What was the weather like – cold/snowy/icy?
- Is the patient unsteady on feet – any risk of falling?
- Did the patient have large amounts of money with them?
- How did they get help?

Previous episodes:

- When?
- How many times?

- What happened on previous occasions?

- Is there any pattern to these episodes?

Causes of wandering:

- Organic cause like dementia, delirium

- Depression

- Psychotic symptoms

- Searching for someone or something

- Loneliness/boredom

- Disorientation to time

Current risks:

- Does she live alone?

- What sort of accommodation?

- How does she manage housework, laundry, shopping, cooking?/Does she need any assistance?

- Is she taking her medication as she should or taking too much?

- Does she go out socially? Is she lonely?

- Does she have any mobility problems?

- Does she drive?

- How does she manage her finances?

- Other risks at home- falling, leaving cooker/fire on, leaving gas on unlit, risk of fire if she smokes
- Does she invite strangers into her home, does she use the security door chain, does she know how to seek help in case of emergency?
- Has she got a panic alarm
- Is she drinking alcohol excessively

Management:

- Assess the current level of support and MDT assessment
- Reminders in the home eg large clock, calendar, note on door, don't leave house on your own, alert helpful neighbours
- Pharmacological- treat depression and psychosis
- Consider acetylcholinesterase inhibitors if not already prescribed
- Residential care may be necessary as wandering may still continue despite other measures to manage risks
- Home with increased package of care – daily carers, meals on wheels, day care, OT assessment – kitchen and fire assessment, smoke alarm, CPN
- Sheltered housing – independent living with on site warden during working hours
- Very sheltered housing – independent living but warden present 24/7

- Elderly mentally ill residential – own room in residential home with staff available 24/7. Meals and help with self care provided for moderate to severe dementia

- Elderly mentally ill nursing- similar to EMI residential but with provision to cope with physical health needs

EXPLAIN TEMPORAL LOBE EPILEPSY

Triggers:

- Change in medications

- Drugs and alcohol

- Stress

- Lack of sleep

Aura:

- Autonomic or visceral sensations (anxiety, epigastric sensations, nausea)

- Funny taste or smell

- Forced thinking

- Déjà vu- do you get feelings of familiarity with circumstances that have not been experienced before?

- Jamais vu- do you get feelings of unfamiliarity with circumstances that you have experienced before?

- Perceptual experiences- hallucinations, illusions, macropsia

Seizure:

- Onset, type- absence/complex partial, frequency
- Absences – lowering of consciousness during ictal phase, less than 5 mins, able to hear events but unable to respond

Automatisms: (ictal phase)

- Lip smacking, finger rolling, throat clearing, bizzare movements
- Dazed state- does not rememeber anything

Post seizure:

- Confusion, tiredness, paranoia
- Post ictal psychosis

History:

- Febrile seizures as a child
- Previous history of seizures
- Mood disturbances

Driving:

Advise not to drive and inform that they have the legal duty to inform DVLA (cannot drive for 1 year)

POST HEAD INJURY ASSESSMENT- HISTORY

Site and severity of head injury

Type of injury, period of unconsciousness, length of post traumatic amnesia, reports of any CT/MRI scans completed at the time

Physical sequalae - hemiparesis, epilepsy

Cognitive sequalae - anterograde/retrograde memory problems, problems with concentration and motivation, intelligence, abstract abilities, judgement and reasoning,

Behavioural problems- social behaviour and disinhibition, personality change, impulsivity

Self care

Rule out other psychiatric conditions – depression, PTSD, psychosis

Premorbid personality and functioning

Previous psychiatric illness

Risks – driving, danger due to aggression, vulnerable to exploitation by others, compliance with medications

POST HEAD INJURY ASSESSMENT OF COGNITIVE FUNCTION (MAINLY FRONTAL LOBE)

This is one of the commonest stations that candidates fail, even though it appears straight forward. You are dealing with a (presumably) frontal lobe impairment patient and hence introduction and rapport is important here. Explain that you are there to assess him because his family is concerned regarding some of the problems he might be having. Explain that you will be examining his front part of the brain which deals with his thinking and memory. Reassure him that, even though some questions may sound unusual, if he gets it wrong not to worry about it.

Start with asking him an open question whether he has expereineced any problems, any change in personality and then ask basic MMSE questions like orienattion. Qucikly move to frontal lobe tests. When you do these tests, remember their frontal lobe is affected - speak slowly checking regularly whether they understand your instructions. It is not simply about completeing your task!

Verbal fluency:

FAS test- Could you tell me as many words beginning with the letter F in one minute (don't count repetitions or proper nouns)

Tell him that you would repeat the same with the letters A and S, at the end of the task, if there is time (>15 words is normal)

Abstract reasoning:

Proverb interpretation– people in glass houses shouldn't throw stones

Cognitive Estimates:

Height of the building, distance from Sheffield to London

Abstract similarities:

Ask the patient in what way are the following similar- chair and table, banana and orange, tulip, rose and daisy

Motor sequencing (Luria's test):

Hand sequence –(fist – edge – palm) do 3 times with the patient and then ask them to do it alone 6 times

Sequence

Alternate sequence

Go-no-go test:

When I tap once you tap twice, when I tap twice you tap once

When I tap once you tap twice, when I tap twice you do nothing

Frontal release signs:

Grasp reflex- You gently inserted your hand into the patients hand and stroke the palmar surface from proximal to distal. Positive response is when patient grasps your hand and continues to grasp as you move your hand.

Palmomental reflex- Draw your finger from the thenar eminence at the wrist up to the base of the thumb. Positive response is when the skin over the chin wrinkles and the corner of the mouth elevates on the same side.

KORSAKOFF'S ASSESSMENT

Start by asking history of memory problems.

Ask about Wernicke's encephalopathy – confusion, ataxia, opthalmoplegia/nystagmus.

Demonstrate chronic alcohol consumption

Test orientation

Test recent and remote memory - Confabulation

Test immediate memory- can't form new memories, repeat questions, can't remember things after immediate events

Test temporal sequence of events

Frontal lobe testing – verbal fluency, abstraction,cognitive estimates,digit span,

Coordination, visual fields

Notes to remember:

Other causes – poor nutrition, gastric cancer, haemodialysis, hyperemesis gravidarum, gastric plication, carbon monoxide poisoning, basal/temporal encephalitis

Investigations – bloods tests to rule out acute infection – FBC, U+E, CRP, chest x-ray, CT head to exclude pathology such as CVA or subdural haematoma

Treat cause of confusion. 80% will never recover.

Once acute confusion settled, investigate underlying chronic condition.

MANAGEMENT OF CHALLENGING BEHAVIOUR IN DEMENTIA

Non pharmacological:

Posters in the house, OT home assessment, organizing shopping trips, support workers

Pharmacological:

Used only in severe distress or risk of harm to self/others

Mild/moderate – No antipsychotics

Severe- Can be considered after risk/benefit analysis

- Assess risk factors and co-morbidity

- Low dose, titrated slowly, short period

- Monitor worsening EPSEs

Acetylcholinesterase inhibitor can be used in Dementia of Lewy Body and in mild-moderate or severe AD if non pharmacological approach has failed

Cautiously use Lorazepam, Haloperidol or Olanzapine

NON-EPILEPTIC SEIZURES- HISTORY

Before the attack:

- Aura

- Any stress (look for emotional precipitant with evidence of secondary gain)

- Psychiatric history

- Family history of epilepsy

- Personal history (childhood abuse)

- Relationship history

- Premorbid personality mainly hysterical personality traits

Attack:

- Sudden or gradual onset

- Wide range of bizarre events like talking, screaming, struggling

- Pelvic thrusting is characteristic

- Lack of tonic/clonic, tongue biting, incontinence and injuries

- Occurs only when other people are around

After attack:

- Absence of post-ictal confusion

- Apparent indifference to the current circumstances or clear evidence of other hysterical behaviour during the interview

Rule out depression, anxiety, psychotic disorders

Task of introducing psychological aetiology for symptoms should be done sensitively without appearing to dismiss the impact of these symptoms

NON EPILEPTIC SEIZURES (NES)- INFORMATION TO CARER

What are NES?

NES are used to describe seizures that often look like epileptic seizures but which have a different cause. Unlike epileptic seizures, they are not caused by changes in brain activity.

What are seizures?

A seizure is a suddent short event where there is a change in the person's awareness of where they are or what they are doing, Causes include epilepsy, alcohol withdrawal, blood sugar changes (low or high), faints, infections

What are the causes of NES?

As described above it could be physiological causes or psychogenic in nature due to stressful psychological experiences or emotional trauma. This is one way that the body indicates excessive stress. NES is often associated with past trauma, childhood abuse, previous contact with mental health services and evidence of past maladaptive coping under stress.

Are they faking?

No. It is very important to recognise that these seizures are real events, although they are different from epileptic seizures. In the past they were sometimes thought to be making it up or attention seeking. We now know that this is not the case and it is important to diagnose them correctly so that people who have NES can get appropriate treatment and not falsely traeted with antiepileptics.

How common is it?

They are common in epileptic population (20% of those with epilepsy) , less common in general population (30 in 100,000)

What are the investigations?

- EEG- normal during and after the attack
- Serum prolactin level increases in epilepsy, normal in NES (although can somewhat raise within 20 minutes of the seizure)
- From the history it has to be differentiated from factitious disorder and malingering

How do you treat it?

Sensitive non-judgemental provision of the diagnosis to patient and family. In 50-70%, they become seizure free after this.Explain to the patient that it is real and that we do not think that they are putting it on.

Psychoeducation about possible aetiological and maintaining factors.

Family based CBT

Individual supportive psychotherapy may bolster resilience and reduce potential for relapse

Longer term – referral for dynamic psychotherapy or local specialist personality disorder service may well be indicated.

Adjunctive use of medication to treat co-morbidity.

MINI MENTAL STATE EXAMINATION (MMSE)

Orientation- year, season, date, day, month (5)

Orientation- country, county, town/city, building, floor (5)

Registration- name 3 objects (apple, table, penny). Ask patient to repeat all 3 after you have said them (3 trials maximum)   (3)

Attention- spell WORLD backwards (5)

Recall- of three objects above (3)

Language- name pencil and watch (2)

Repeating- Repeat "no ifs ands or buts" (1)

Three stage command- take paper in your right hand fold it in half and put in on the floor (3)

Reading- Read and obey the following (1)

CLOSE YOUR EYES

Write a meaningful sentence (1)

Copy pattern (1)

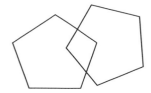

PARIETAL LOBE TESTING

Finger agnosia- touch nose with index finger

Astereognosia- recognise coins in the hand

Dysgraphaesthesia- write numbers on hand

Asomatognosia- recognise parts of body

Repetition – no ifs ands or buts

Read – CLOSE YOUR EYES

Write a sentence

Constructional dyspraxia

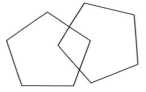

Test visual fields

BREAKING BAD NEWS

Summarise symptoms patient presented with

Discover what happened since last seen (any change in symptoms)

Get an idea of how patient/carer are thinking/feeling

Assess patients understanding first

Warning shot that bad information is coming

Relate your explanation to the patient/carers understanding

Dont give too much information too early – small chunks – categorise information

Watch the pace and check repeatedly for understanding

Allow for patient to shut down – give space and allow possible denial

Keep pausing to give patient chance to speak

Check understanding – would you like to run through what you will tell your wife?

Be aware of unshared meaning (i.e. what diagnosis means to the patient/carer)

Get her opinion

Mention about specialist treatment and follow up, Macmillian nurses and palliative care

CHAPTER 4   ALCOHOL AND DRUGS

DEPENDENCE SYNDROME (three of the following in the previous year)

- Strong desire to take the drug

- Difficulties in controlling its use

- Persisting in its use despite harmful consequences

- Higher priority given to drug use than to other activities and obligations

- Increased tolerance

- Physical withdrawal state (tremor, sweating, nausea, retching or vomiting, tachycardia or hypertension, psychomotor agitation, headache, anxiety, depression, insominia, malaise or weakness, transient auditory, tactile or visual hallucinations, grand mal convulsions)

ELICIT HISTORY OF ALCOHOL DEPENDENCE

- Life time pattern of alcohol consumption

    - Age at first drinking

    - Age at regular drinking

    - Age at drinking most days

    - Periods of abstinence

- Current alcohol consumption

- Signs of dependence

- CAGE questionnaire (cut down on drinking, getting annoyed when people discuss about it, feeling guilty about it, drinking first thing in the morning- eye opener)

- Complications

  - Medical- liver problems

  - Accidents/ head injury

  - Memory problems

  - Black outs

- Social problems

  - Work

  - Relationships

  - Family

  - Financial

- Forensic history

- Co-morbidity

  - Affective symptoms (including suicidal ideations)

  - Psychosis

  - Anxiety symptoms

- Assess insight and motivation

- Past psychiatric history

- Family history

ALCOHOL WITHDRAWAL- MANAGEMENT

Alcohol withdrawal symptoms ususaly seen within 3-12 hours and peak in 24-36hours, Epilieptic seizures most commonly occur 12-18 hours after the last drink.

- Admit the patient

- FBC, U + Es, LFTs, TFTs, CRP

- CT/MRI (if required)

- EEG if history of blackouts

- MMSE as screening

	Biological	Psychological	Social
Immediate	BDZ detox (Chlordiazepoxide Oxazepam if liver disease), add PRN if required. Parenteral thiamine Sodium valproate if at risk of seizures. Monitor diet, fluids, input, output, BP, P, Temp	Supportive psychotherapy. Close nursing care for suicide. *prevention* Reality orientation Reassurance	Assess social circumstances.

	Biological	Psychological	Social
Short Term	Multivitamins Assess for affective and psychotic symptoms	Motivational interviewing. Group therapy	Practical problem solving
Longer Term	Acamprosate (anti craving) Disulfiram (Aversion therapy)	Relapse prevention Day/residential rehab	AA Social skills Vocational training Halfway house Al anon

Watch for Wernicke's encephalopathy and Korsakoff's.

Hallucinations – oral haloperidol

Acamprosate

- 666mg TDS
- Side effects- GI upset, pruritus, rash, altered libido
- No addictive potential

Disulfiram

- Don't take if heart problems, heart failure or diabetes.
- Has to be supervised.

- If alcohol is consumed- flushing, nausea, vomiting, tachycardia

- 200-400mg on 3 days of the week

- Side effects- halitosis, headache, rarely hepatoxicity and psychotic reaction

MOTIVATIONAL INTERVIEWING

Intention to change:

- What are your worries/ difficulties about the current situation?

- what do others think/

- Why does you think you need to change?

- What needs to change?

- How important is it?

- How motivated are you?

- What do you think you might do to achieve this?

Advantages of current situation:

- What do you like about drinking?

- What are the advantages of maintaining your current situation?

Disadvantages of current situation:

- What do you think will happen if you do not change?

Advantages of changing:

- What would be good about stopping drinking?

- Do you think your physical/mental health will improve?

Summarise

SELF HELP

- Set a target of how much you want to reduce by

- Keep a diary to cut down

- Avoid high-risk drinking situations

- Drink lower-strength drinks

- Do other activities instead of drinking.

- Involve your partner/friends/family so they can help you achieve your goals

- See GP for advice on cutting down

- May need medications (Librium) for a short time

- Join support groups – Alcoholic Anonymous groups

OPIATE USE AND PREGNANCY

- Effects:

    - 1st trimester-intrauterine growth retardation, low birthweight, miscarriages

    - 3rd trimester-preterm delivery, sudden infant death syndrome, neonatal abstinence syndrome, still birth

- Opiate detox usually is suggested in the second trimester

- Opiate replacement therapy- Methadone (Buprenorphine can be used if not pregnant)
  - Supervised by pharmacy
  - Methadone is long acting, therefore more stable for mother and baby
- Becoming drug free prior to birth of baby risks accidental overdose if she uses methadone to cope with stress of pregnancy
- Harm minimization strategies- need exchange, safer injecting techniques
- Psychosocial interventions- motivational interviewing/Narcotics Anonymous
- Treat co-morbid psychiatric problems
- Involve child and family services

COCAINE AND PREGNANCY

- Intrauterine growth retardation, low birth weight, foetal death, premature delivery, foetal brain infarctions from cocaine binges

TOBACCO AND PREGNANCY

- Three times more likely to have low birth weight
- Still births and neonatal deaths are 30% higher
- Advise stop smoking or cut down
- Nicotine Replacement Therapy not licensed in UK for pregnancy

CANNABIS AND MENTAL HEALTH

- Can be used as resin, herbal cannabis, skunk (higher concentration of active ingredient)

- Can be detected up to 56 days after it has been smoked

- Class B drug – 5 years + unlimited fine for possession, 14 years + unlimited fine for supplying

- Gives feeling of being high, colours and sound more intense and sleepiness

- 1 in 10 have confusion, hallucinations, anxiety and paranoia

- Reduced motivation and therefore has depressant effect, decreased concentration and ability to organise information – affecting education

- Regular use doubles the risk of developing psychosis or schizophrenia. There is a clear link between cannabis use and psychosis in those who have a genetic susceptibility

- Higher risk of relapse in schizophrenia. If you started smoking before 15, youre 4 times more likely to develop schizophrenia by age 26 according to research. Teenagers are more susceptible due to developing brain

- Cannabis psychosis – short lived psychotic illness brought on by use of cannabis which subisides fairly quickly once individual stops

- It has some addictive features – tolerance, withdrawal (craving, decreased appetitie, sleep difficulty, weight loss, aggression, irritability, restlessness) compulsive use and even dependence

- TALK TO FRANK website, drug and alcohol team

70

CHAPTER 5   SCHIZOPHRENIA

SCHIZOPHRENIA

a) Thought echo, insertion, broadcast and withdrawal

b) Delusions of control, influence, or passivity of body or limbs, clearly referrred to body or limb movements or specific thoughts, actions or sensations, delusional perception

c) Hallucinatory voices giving a running commentary on the patient's behaviour or discussing the patient among themselves, or other types of hallucinatory voices coming from some part of the body

d) Persistent bizarre delusions

e) Persistent hallucinations in any modality

f) Breaks in thought, resulting in incoherence or irrelevant speech; neologisms

g) Catatonic behaviour- excitement, posturing, waxy flexibility, negativism, mutism and stupor

h) Negative symptoms – apathy, paucity of speech, blunted emotions, social withdrawal

i) Significant and consistent change in the overall quality of some aspects of personal behaviour- loss of interest, aimlessness, idleness, social withdrawal, self absorbed attitude.

One of a-d (and usually two or more if less clear cut) for 1 month or more or at least two of e-h

PARANOID SCHIZOPHRENIA

- This is dominated by relatively stable, often paranoid delusions, usually accompanied by hallucinations, particularly of the auditory variety, and perceptual disturbances.

- Disturbances of affect, volition and speech, and catatonic symptoms, are either absent or relatively inconspicuous.

HEBEPHRENIC SCHIZOPHRENIA

- Affective changes are prominent, delusions and hallucinations fleeting and fragmentary, behavior irresponsible and unpredictable, and mannerisms common.

- The mood is shallow and inappropriate, thought is disorganized, and speech is incoherent. There is a tendency to social isolation. Usually the prognosis is poor because of the rapid development of "negative" symptoms, particularly flattening of affect and loss of volition.

- It should normally be diagnosed only in adolescents or young adults.

- Have to be observed for 2-3 months for before offering diagnosis

SCHIZOTYPAL DISORDER

Symptoms occur repeatedly over a two year period

a) Inappropriate or constricted affect

b) Odd, eccentric or peculiar behaviour

c) Poor rapport with others and tendency to social withdrawal

d) Odd beliefs and magical thinking

e) Suspiciousness or paranoid ideas

f) Obsessive ruminations without inner resistance, often with dysmorphophobic, sexual or aggressive contents

g) Unusual perceptual experiences

h) Vague, circumstantial, overelaborate, metaphorical or stereotyped thinking manifested by odd speech without gross incoherence

i) Occasional transient quasi psychotic symptoms with intense illusions, auditory or other hallucinations and delusion like ideas

DELUSIONAL DISORDER

- Delusion or set of related delusions other than those that are typically schizophrenic

- Must be present for at least 3 months

- General criteria for schizophrenia are not fulfilled

- No persistent hallucinations in any modality (can be transitory)

- Depressive symptoms may be present intermittently or even a depressive episode, provided delusions persist when there is no disturbance of mood

ACUTE AND TRANSIENT PSYCHOTIC DISORDERS

- Acute onset of psychotic symptoms such as delusions, hallucinations, and perceptual disturbances, and by the severe disruption of ordinary behaviour.

- Acute onset of two weeks or less

- No evidence of organic causation.

- Complete recovery usually occurs within a few months, often within a few weeks or even days. The disorder may or may not be associated with acute stress, defined as usually stressful events preceding the onset by one to two weeks.

SCHIZOPHRENIA- EXPLANATION

What is Schizophrenia?

It is a mental illness where people find it difficult to decide what is real and what is not. They may act strangely, thinking may be muddled and may have abnormal experiences. It affects 1 in 100 people, men and women equally.

What are the symptoms?

They may have unusual experiences like hearing voices when there isn't anybody around and holding odd beliefs. They may have muddled thinking making it harder to concentrate, thoughts seem to wander, drift from idea to idea without any obvious connection between them and feelings of being controlled.

They may have poor concentration, difficulty looking after themselves, sitting still, gaze blankly, decreased facial expression and eye contact, monotonous speech, lack of motivation, decreased enjoyment, decreased physical activity, relationship difficulties- decreased friendships, isolative, decreased intimacy and social interest.

What causes schizophrenia?

There is no one single cause.

- **Genes:** If there is a family history of schizophrenia, the children are more at risk- 1 in 10 has a parent with schizophrenia.

- **Chemical imbalance**: There is chemical imbalance of dopamine and serotonin.

- **Brain damage**: Brain damage in early life or birth leading to abnormal brain development– problems with oxygen supply to the brain at birth, viral infections during early months of pregnancy

- **Street drugs and alcohol**: Drugs like cannabis (doubles the risk of schizophrenia, and if they started to use regularly in their teens, then they are 4 times more likely to get schizophrenia before the age of 26), cocaine, amphetamines, LSD and ecstasy can bring on schizophrenia. Drinking excessive amounts of alcohol can bring about symptoms similar to schizophrenia.

- **Stress**: Stress can predispose to first onset psychosis and sometimes cause relapse.

- **Family problems**: At one time it was thought that communication problems within the family causes schizophrenia. We now know that there is no evidence to support this. However family tensions can make things worse.

- **Childhood deprivation**: Early experiences of deprivation and abuse can make it more likely that they will develop schizophrenia.

Are people with schizophrenia dangerous?

People who have schizophrenia are not usually dangerous. Violent behaviour is usually caused by street drugs or alcohol, which is similar to people who don't suffer from schizophrenia.

Is schizophrenia a split personality?

This is not true. People with schizophrenia have only one personality, although their personality may be disturbed in some way.

Why is treatment important?

Suicide is more common with schizophrenia. Research suggests that longer the schizophrenia is untreated, greater is its impact on life and sooner it is treated, the better the outlook.

What treatments are available?

Biological, psychological and social

Biological (medications)

The aim is to reduce the effects of the symptoms on their life and medications weaken the odd beliefs and the unusual experiences, help them to think clearly, increases the motivation and the ability to look after themselves and also reduces the chances of relapse.

They are treated with antipsychotics which restore chemical imbalance in the brain involving dopamine. There are two types, typical (older) antipsychotics and the atypical (newer) antipsychotics. Older antispychotics include medications like Haloperidol and Chlorpromazine. Side effects include stiffness, shakiness, restlessness, problems with sex life (in 45% - decreased libido, disorder or arousal, erection and ejaculation). Longer time side effects include persistent movement of the mouth and tongue (tardive dyskinesia) which affects about 1 in 20 people who take these medications.

Newer antipsychotics work on a different range of chemicals in the brain (including serotonin). They are less like to cause abnormal body movements or tardive dyskinesia, although with high doses they can occur. They also help with some of the negative symptoms like motivation and concentration. Some of the side effects include sleepiness, slowness, weight gain (mainly in1st 6 months then settles), interference with sex life, increased chance of developing diabetes.

The medications may cause initial drowsiness so don't operate machinery or drive. They can drink alcohol in moderation, although may make them more sleepy. Inform the insurance company if they plan to drive.

Longer acting depot preparations available if compliance is a problem. This is a slow release form of the antipsychotic medication given as an injection, ranging from once a week to once a month. It is usually given by a nurse. A test dose is usually given to see whether they develop side effects. If they don't experience any side effects, 5-10 days later they would be given the full dose. Side effects include pain at injection site, restlessness, drowsiness, dizziness, constipation, dry mouth, blurred vision, weight gain, movement disorders and sexual problems.

Clozapine?

- This is one of the newer medications and is more effective for people who do not respond to two antipsychotics. It also decreases sucidality and improves tardive dyskinesia. It acts on dopamine.

- Prior to starting Clozapine, you have to register with the official monitoring agency (CPMS). Full blood investigations and ECG will be done.

- Side effects include dizziness, drowsiness, fluttering of the heart, extra saliva, weight gain, constipation, increased risk of diabetes.

- Rare but serious side effect- can affect your bone marrow. This leads to a shortage of white cells which makes you vulnerable to infection. If this happen, the medication needs to be stopped as quickly as possible to allow the bone marrow to recover (neutropenia- 1%, agranulocytosis- 3%). Hence it is important to let the doctor know if you get sore throat or fever. Because of this blood tests will be done weekly for the first 18 weeks, then fortnightly up to one year and then monthly thereafter.

- You will be started on a small dose and then increased slowly. If you miss the dose for more than 48 hours, it will be restarted from the starting dose.

- Levels of drug can be affected if you smoke. Avoid alcohol as it may make you more sleepy. There is interaction with other medications like cotrimoxazole, chloramphenicol and carbamazepine (doesn't mean you cant take it, just follow your doctor's instructions)

- 6 out of 10 get better on clozapine.

- In pregnancy, the safety is not clear. It is secreted in breast milk and must be avoided.

How long should the medications be taken?

1-2 years for acute episode, reviewed every 3-6 months. It is important to continue even if you feel well, as stopping doubles the risk of relapse.

What is the prognosis?

- 1 in 5 will get better within 5 years of first episode

- 3 in 5 will get better but have residual symptoms

- 1 in 5 will continue to have troublesome symptoms

Psychological treatment

- Cognitive behavioural therapy can help them to live with their experiences or help to work out what makes them unwell. It looks at problems and how they think, feel and behave in relation to these.

- Counselling is provided if they need to talk to someone or if they need support with the problems of daily life.

- Family work- Education with family can help the family to learn about the disorder, ways to support them and help to solve practical problems.

Social treatment

- Day centres

- Work projects (to improve skills)

- Supported accommodation

- Occupational therapy

- CPN

- CMHT

- CRHT for crisis management

Self help

- Learn to recognise when you are becoming unwell

- Beware of the signs of relpase

- Avoid things that worsen- stressful situations, getting anxious about bills, arguments, drugs and alcohol

- Learn relaxation techniques

- Do something you enjoy regularly

- Keep busy to control voices- spend time with other people, personal stereo

- Remind yourself that voices can't harm you and don't have any power over you

- Join a self-help group

- Identify someone you can trust to tell if you get unwell

- Learn about schizophrenia and medication

- Look after your body- balanced diet, exercise, avoiding smoking

How can the family help?

- Handling positive symptoms (dont be shocked, tactfully change the subject but dont argue with him) and negative symptoms (encourage him, positive praise, don't force him, encourage daily routine)

- Suicidal symptoms – contact care team

- Don't pressurize

- Lower expectations – dont expect him to perform the same as before he was unwell

- Encouraging daily structure and routine

- Encourage compliance with treatment and to discuss and side effects

- Study early warning signs

- Looking after their own health – carer's assessment

- Find out more about the illness

- Attend carers groups

PARAPHRENIA

- Check hallucinations in all sensory modalities- olfactory, gustatory, somatic, auditory and visual

- Aetiology:

  - genetic factors- risk of schizophrenia in 1st degree relatives is 3.4%

  - Premorbid personality – poor adjustment and 45% show paranoid/schizoid traits

  - Sensory impairments – hearing/vision

  - Social isolation

  - Substance misuse especially alcohol

- Persecutory delusions commonest in 90%, auditory hallucinations in 75% and visual hallucinations in 13%.

- Negative symptoms, catatonia and thought disorder are uncommon.

- Treated by relieving sensory deficits and low dose atypical antipsychotics

DELUSIONAL JEALOUSY- HISTORY AND MANAGEMENT

- Index Episode and relationship history (and past relationships)

- Evidence for partner's infidelity

- Check for strength of belief- How strongly do you believe this?/Is it a possibility that you could have misunderstood?

- Any other abnormal beliefs- Paranoid/persecutory idea/delusion/other psychotic symptoms

- Anxiety symptoms

- Past psychiatric history

- Drug/ alcohol history

- Medical history (rule out organic cause)

- Forensic history

- Premorbid personality (always been jealous)

- Risk assessment:

  - wife

  - self

- others

- access to weapons

- Management:

    - In the community/ admission to hospital (use of the Mental Health Act)

    - Geographical separation

    - Antipsychotics

    - Psychological therapy- Individual therapy, supportive psychotherapy, insight oriented therapy

EROTOMANIA

The history taking approach is similar to delusional jealousy. In addition:

- Safe assessment- confirm weapons have been given to staff

- History of offending

- History of violence to others

- History of stalking

- Risk assessment – knife and reason for carrying, risk related to staff member, risk to others

- Management:

    - Discuss nature of the problem- patient's feelings toward him, contents of the bag, risk to potential victim and family and other risk factors

    - Needs to inform police and get legal advice through the trust

    - Inform consultant, ward manager, security at the hospital

- Explain stalking and risks (willful, malicious and repeated following or harassing of another person that threatens his or her safety". Usually includes the intrusive following of a 'target'. Stalkers most often persecute their targets by unwanted communications, like frequent telephone calls, letters, e-mail, notes or packages)

- MDT approach

CHAPTER 6   MOOD DISORDERS

HYPOMANIA

- Persistent mild elevation of mood, increased energy and activity, and usually marked feelings of well-being and both physical and mental efficiency.

- Increased sociability, talkativeness, over-familiarity, increased sexual energy, and a decreased need for sleep are often present but not to the extent that they lead to severe disruption of work or result in social rejection.

- Not accompanied by hallucinations or delusions

MANIA WITHOUT PSYCHOTIC SYMPTOMS

- Mood is elevated out of keeping with the patient's circumstances and may vary from carefree joviality to almost uncontrollable excitement.

- Elation is accompanied by increased energy, resulting in over activity, pressure of speech, and a decreased need for sleep.

- Attention cannot be sustained, and there is often marked distractibility.

- Self-esteem is often inflated with grandiose ideas and overconfidence.

- Loss of normal social inhibitions may result in behavior that is reckless, or inappropriate to the circumstances, and out of character.

- For at least 1 week

MANIA WITH PSYCHOTIC SYMPTOMS

In addition to the above clinical picture, delusions (usually grandiose) or hallucinations (usually of voices speaking directly to the patient) are present, or the excitement, excessive motor activity, and flight of ideas are so extreme that the subject is incomprehensible or inaccessible to ordinary communication.

BIPOLAR AFFECTIVE DISORDER

Characterized by two or more episodes in which the patient's mood and activity levels are significantly disturbed, this disturbance consisting on some occasions of an elevation of mood and increased energy and activity (hypomania or mania) and on others of a lowering of mood and decreased energy and activity (depression).

BIPOLAR AFFECTIVE DISORDER (BPAD)- EXPLANATION

What is BPAD?

- Used to be called manic depression
- Characterised by mood swings- can be low (feeling depressed and unhappy) or high (happy and elated, full of energy, not wanting to eat, problems with sleep, disinhibited and overactive) or mixed (depressed mood with the restlessness and overactivity of a manic episode)
- Might also develop psychotic symptoms like feeling uniquely guilty or that you have special powers when manic.

- Affects 1:100 adults, men and women equally

- Usually starts before the age of 30

Causes?

There is no single cause. Possible causes include

- **Runs in families** – more to do with genes than upbringing

- **Chemical imbalance**

- **Stress**- risk of precipitating episodes

Rapid cycling?

More than four mood swings happen in a 12 month period. This affects around 1 in 10 people with bipolar disorder.

Cyclothymia?

The mood swings are longer but less severe and never meet the criteria for depressive or manic/hypomanic episodes

Treatment?

Biological (Mood stabilizers- Lithium, Semi sodium valproate, Olanzapine)

Psychological

Social

LITHIUM

- Mood stabilizer

- Leads to fewer manic episodes and reduces relapse rate by 30-40%

- Mechanism – unclear- may correct chemical imbalance.

- Prior to starting- FBC, ECG, kidney and thyroid function test.

- Taken as a tablet

- Usual dose 400-600mg at night, increased weekly max 2gram per day. Check levels 5 days after starting and 5 days after each dose change. Blood samples should be 12 hours post dose. Treatment level 0.6-1.2mmol/L. (>1.5 toxic)

- Lithium and U & Es checked every 3 months, TFTs every 6-12 months and creatinine clearance every 12 months

- Side effects:

  - Most common- feeling thirsty, passing more urine than usual, metallic taste in the mouth, weight gain (about 25% of patients gain more than 4.5 kgs)

  - Less common- blurred slight muscle weakness, occasional diarrhoea, fine hand tremor, feeling of mildly ill, cardiac conduction problems

- Rare- sexual side effects (impairment of drive, arousal and ejaculation)

- Long term- Kidney changes in 10-20%, hypothyroidism in 5-35% (generally reversible on discontinuation)

- Toxicity- persistent diarrhoea, vomiting, severe hand tremors, slurred speech, confusion, lack of coordination, dehydration, lethargy, drowsiness, collapse, seizures

- Amount of Lithium in the blood is very sensitive to how much or how little water is there in the body. If you become dehydrated it can lead to toxixity. Hence you need to maintain water balance, avoid excess caffeine and avoid too much activity in hot weather.

- Use appropriate contraception as it can damage the baby

- May impair co-ordination so need to be careful while driving and operating machinery.

- Safe with alcohol in moderation

LITHIUM AND PREGNANCY

- Ebstein' s anomaly- 8 fold relative risk in the first trimester

- Congental anomalies- 4-12%

- Risk vs benefit

- Mild BPAD– reduce and discontinue pre pregnancy

- Moderate BPAD- taper and discontinue in 1st trimester

- Severe BPAD- maintain lithium with informed consent and appropriate counseling, Echocardiogram at 16-18 weeks.
- It can cause floppy baby
- It is secreted in breast milk. advisable to avoid breast feeding

SEMISODIUM VALPROATE

- Used in the treatment of BPAD to prevent extreme mood swings.
- Works on a chemical in the brain called GABA preventing its breakdown (GABA prevents fits and overactivity)
- Blood test in the first 6 months to check liver function and full blood count.
- Need to take for several months to years.
- Side effects- Drowsiness, nausea, increased appetite, weight gain, hair loss, oedema, may affect periods, skin rashes, changes in blood count , tremor, gait disturbance. May feel drowsy initially, no effect on sex life; very rarely-pancreatitis, liver failure

CARBAMAZEPINE

- It works by attaching to channels that control sodium getting into the cell and affects them
- Side effects- nausea, vomiting, diarrhea, constipation, blurred vision, feeling tired or dizzy, 10% develop mild rash, 1 in 200 develop serious rash requiring urgent treatment; can affect heart conduction, kidneys and cause hepatitis

- Toxicity- dizziness, ataxia, sedation, diplopia

- FBC & LFTs every two weeks for the first two months and then every 3 months.

- May interact with oral contraceptive pill

- Can affect unborn baby causing spina bifida, facial defects

- Avoid breast feeding

LAMOTRIGINE

- Used to prevent severe depressive episodes.

- Can cause Stevens Johnson syndrome

ANTIPSYCHOTICS

Olanzapine, Quetiapine, Risperidone

PSYCHOLOGICAL TREATMENTS

These can be particularly helpful in between episodes of mania or depression.

- Psychoeducation – finding out more about bipolar disorder

- Mood monitoring – picking up mood changes

- Mood strategies – preventing mood from going to full manic or depressive episode
- Developing general coping skills
- Cognitive behavioural therapy (CBT) for depression.

How long should the mood stabilizer be taken for?

Continue for 2 years after 1st episode, for up to 5 years if there are risk factors such as frequent relapses, drug/alcohol problems, continuing social stress, psychotic episodes.

Course?

- 80% chance of further episodes
- Second manic episode usually within 2 years of the first.
- 7% have only one episode

Driving?

Have to stop when unwell. To restart driving you have to be stable for 3 months (6 months if rapid cycling), compliant with treatment, having no side effects that would impair driving, having full insight and subject to specialist report.

SELF HELP

- Realise when your mood is getting out of control

- Educate yourself about the disorder

- Avoid anything that causes you stress

- Have someone who you can talk to and confide in

- Make sure there is a good work/life balance

- Regular exercise and balanced healthy diet

- Don't suddently discontinue meds

FOR CARERS/FAMILY/FRIENDS

- Listen and try and understand when person is depressed

- During mania, the person will usually be happy, outgoing and full of energy. Some social situations may push their energy even higher so try and keep them away from parties or heated discussions and advise them they need help.

- Give yourself time and space to have a break too

- Educate yourself about the disorder and maybe even attend appointments with them

NICE GUIDELINES- TREATMENT OF BPAD

- Mild depression- watchful waiting and reassess 2 weeks

- Moderate or severe depression- SSRI or add quetiapine (as long as taking non antipsychotic mood stabilizer)- if no improvement consider CBT

- If poor response to treatment consider increasing antidepressant dose to maximum, adding mirtazapine or venlafaxine, adding quetiapine or olanzapine, or adding lithium. If non-responsive to three courses of antidepressant refer to specialist unit.

- Concurrent depressive and psychotic symptoms- augment with antipsychotic like olanzapine, quetiapine or rispseridone

- ECT for severe depressive illness, prolonged or severe mania, catatonia

- Rapid cycling
  - Acute episodes- as per manic and depressive episodes
  - Long term- 1st line use lithium and valproate, 2nd line use lithium monotherapy, combining lithium or valproate with lamotrigine
  - Avoid using antidepressants

- Mixed episodes- treat as if they had an acute manic episode. Avoid antidepressants

DEPRESSIVE EPISODE

In typical mild, moderate, or severe depressive episodes, the patient suffers from depressed mood, loss of interest and enjoyment (anhedonia), and reduced energy leading to increased fatiguability and diminished activity. Marked tiredness after only slight effort is common.

Other common symptoms:

- Reduced concentration and attention
- Reduced self esteem and confidence
- Ideas of guilt and unworthiness
- Bleak and pessimistic views of the future
- Ideas or acts of self harm or suicide
- Disturbed sleep
- Diminished appetite

Duration of at least 2 weeks is usually required for a diagnosis, but shorter periods may be reasonable if symptoms are unusually severe and of rapid onset.

Mild depressive episode (2+2)

- At least 2 of depressed mood, anhedonia, decreased energy/fatiguability
- Plus at least 2 of the other symtoms

Moderate depressive episode (2+3)

- At least 2 of depressed mood, anhedonia, decreased energy/fatiguability

- Plus at least 3 of the other symtoms

Severe depressive episode with/ without psychotic symptoms (3+4)

- All 3 of depressed mood, anhedonia, decreased energy/fatiguability

- Plus at least 4 of the other symtoms, some of which should be of severe intensity

Somatic syndrome (always present in severe depressive episode)

- Anhedonia

- Lack of emotional reactivity

- Early morning wakening

- Depression worse in morning

- Psychomotor retardation or agitation

- Marked loss of appetite

- Weight loss (5% or more of body weight in last month)

- Marked loss of libido

Recurrent depressive disorder

- Repeated episodes of depression as described for depressive episode without any history of independent episodes of mood elevation and increased energy (mania).

- At least two episodes should have lasted a minimum of 2 weeks and should have been separated by several months without significant mood disturbance

Cyclothymia

Persistent instability of mood, involving numerous periods of depression and mild elation, none of which has been sufficiently severe or prolonged to fulfill the criteria for bipolar affective disorder

Dysthymia

A chronic depression of mood, lasting at least several years, which is not sufficiently severe, or in which individual episodes are not sufficiently prolonged, to justify a diagnosis of severe, moderate, or mild recurrent depressive disorder

DEPRESSION- EXPLANATION

What is depression?

Depression means different for different people. Generally it means extension of feeling miserable. 20% of people will have at some point in their lives. Females more commonly affected.

Symptoms?

As mentioned previously

Causes

There is no single cause.

- **It may run in families**

- **Chemical imbalance** in the brain- serotonin

- **Alcohol** abuse can cause depression

- **Stressful life events**

- **Personality**- making us more vulnerable due to early life experiences

- **Physical illness**

- Certain **medications** like steroids

Treatment?

Can be treated with medications or talking therapies.

Mild depression

- Watchful waiting- review after 2 weeks

- Sleep hygiene, anxiety management, exercise, guided self help for 6-9 weeks
- Computerized CBT
- 6-8 sessions of problem solving therapy, brief CBT and counseling over 10-12 weeks
- No antidepressants unless no response to the above

Antidepressants

- Antidepressants are used to help relieve the symptoms of depression
- Used in moderate to severe depression
- Used for other conditions such as OCD, anxiety, eating disorders (Bulimia) and PTSD. Four main groups SSRIs (selective serotonin reuptake inhibitors), tricyclics, MAOIs (monoamine oxidase inhibitors), SNRIs (serotonin and noradrenaline reuptake inhibitors).
- Works on a chemical imbalance of serotonin which can be decreased in depression. Some medications also work on noradrenaline.
- Taken orally each day
- 6 out of 10 people improve on an antidepressants
- The older medications are just as effective as the newer medications but the newer medications have less side effects and are safer in overdose.
- Not addictive

- Don't stop suddenly due to risk of discontinuation symptoms- dizziness, light headedness, nausea, headache, flu like symptoms, electric shock like sensations, vivid dreams

- Continue for 6 months after 1st episode; 2 years after 2 episodes

SSRIs

- Most commonly used

- In the first 2 or 3 weeks of starting them, you may feel sick and more anxious (prevent the feelings of nausea by taking them with food)

- The side effects usually wear off over a couple of weeks as your body gets used to the medication

- They may interfere with your sexual function. There have been reports of episodes of aggression, although these are rare.

- More serious side effects include- problems with urinating, difficulty in remembering, falls, confusion. These are uncommon in healthy, younger or middle-aged people.

- Inform the doctor if you feel suicidal (can happen sometimes in the initial weeks of treatment)

SNRIs

- Similar side effects to SSRIs.

- Do not take Venlafaxine if you have a serious heart problem

TRICYCLICS

- Side effects include dry mouth, a slight tremor, palpitations, constipation, sleepiness, and weight gain.

- May cause confusion in the elderly, difficulties with starting or stopping when passing water, faintness, and falls.

- Sexual problems include diffulty getting or keeping an erection, or delayed ejaculation.

- Dangerous in overdose.

ST. JOHNS WORT

- Herbal remedy

- Widely used in germany

- Effective in mild to moderate depression

- One tablet per day. Need to consult doctor

- Works in the same way as antidepressants

- Can interfere with other medications such as antidepressants, painkillers, oral contraceptives pill (making it less effective), some cancer drugs. some epilepsy drugs, such as carbamazepine, digoxin, warfarin, HIV drugs.

Psychotherapy

- CBT
- Counselling
- Problem-solving therapy
- Behavioural activation
- Couple therapy- if depression connected to relationship
- Bereavement counselling
- Interpersonal and psychodynamic psychotherapy

SELF HELP

- Do an activity each day- go for a walk or do some exercise
- Talk to someone you trust about how you feel
- Make sure you make yourself eat properly even if you feel you don't want to
- Avoid alcohol as this makes depression worse
- Avoid other drugs too (cannabis in your teens can bring on depression)

- It might be useful to write down what you think is causing your depression and then you may be able to try different strategies to improve things

- Don't lose hope, things will improve

- Self-help books using CBT principles

- Self help computer programmes on internet

FOR CARERS

- Listen and spend time

- Can help the person find solutions to their problem causing their low mood

- Make sure they are eating properly

- Make sure they are avoiding alcohol

- If they are getting worse encourage them to visit their doctor and take the appropriate treatment

DEPRESSION AND SEXUAL SIDE EFFECTS OF MEDICATIONS

- Enquire about current treatment – when started, effects, symptoms improved/not improved, side effects

- Expectations and concerns regarding current and future treatment

- Enquire in detail the reasons for wanting to stop treatment including performance anxiety and sexual dysfunctions
- Enquire psychosexual history sensitively including desire, morning erection and during sex, masturbation, satisfaction
- Treatment options:
  - Allay any anxiety regarding the treatment by reassuring him (if the reason to stop the antidepressant is not severe enough)
  - Explain to the patient that sexual symptoms are common in depression. And also around 5 out of 10 people with depression report problems with libido in the month before the diagnosis. State that spontaneous remission occurs in 10% and partial remission in a further 11%.
  - In the exam do not forget to talk about performance anxiety
  - Consider another drug- reboxetine (lowest risk of sexual side effects), Mirtazapine
  - Medication holidays
  - Viagra (sildenafil) or similar medications

TREATMENT RESISTANT DEPRESSION

Failure to respond to 2 or more antidepressants given sequentially in adequate dosage and for an adequate period of time. Occurs in 10-20% of major depression

- Management

  - Reconsider diagnosis

  - Undiagnosed medical illness like early huntingtons, malignancy

  - Psychological issues- for example, undisclosed sexual abuse, oppressive marital

  relationship

  - Drug treatment adherence – history from informant, presence of side effects

  - Substance misuse

  - Personality disorder

Other treatment options

1) **High dose monotherapy** (high dose venlafaxine >150mg gives advantage over SSRIs)

2) **Add in CBT** if not already done

3) **Switching antidepressants** (usually allow 6-8 weeks)

   Risks- discontinuation symptoms if too quick, withdrawal may exacerbate depression, new side effects from antidepressants and serotonin syndrome

   50% of non responders to 1st antidepressant are responsive to second antidepressant. Avoid switching to dosulepin.

4) **Augmentation with lithium**- enhances serotonin transmission. This requires close monitoring because Lithium toxicity may occur at even normal serum levels; risk of serotonin syndrome; at least 8 weeks trial

5) **Augmentation with antipsychotics (NICE suggests aripiprazole/olanzapine/quetiapine/risperidone)**

6) **Combinations of antidepressants**- avoids risk of discontinuation symtoms, two different pharamacodynamics, built upon partial response to first but there is limited evidence, increased risk of side effects. There is evidence for adding Mirtazapine to SSRIs

7) **ECT**

MODERATE TO SEVERE DEPRESSION IN CHILDREN

- Individual CBT, IPT or shorter term family therapy of at least 3 months duration (If no benefit from 4-6 sessions of psychological therapy try another modality)

- Add fluoxetine 10mg od if 12-18yrs old consider if 5-11yrs old

- If no improvement from further 6 sessions of combined treatment offer more intensive psychotherapy

- Antidepressants only in combination with CBT

- Monitor for any bullying as a contributing factors

- MDT after 4-6 sessions of psychotherapy

- If no response then alternative psychological therapy or systemic family therapy (at least 15 fortnightly sessions), or individual child psychotherapy (approximately 30 weekly sessions)
- Continue antidepressants for 6 months from point of 8 weeks of full functioning if fluoxetine doesn't work consider setraline or citalopram
- Do not use paroxetine, venlafaxine or TCAs, St. Johns Wort

ANTIDEPRESSANTS AND PREGNANCY

- Best to take as little medication as possible in the first three months of pregnancy as may cause congenital malformations
- Some mothers have to take to prevent deterioration and this risk needs to be balanced with that of taking no medication
- At the moment, Fluoxetine seems to be the safest antidepressant to take in pregnancy
- Sometimes babies can get withdrawal symptoms soon after birth- more common with paroxetine

ANTIDEPRESSANTS AND BREASTFEEDING

- When breastfeeding baby is likely to get only a small amount of antidepressant from mother's milk

- Babies older than a few weeks are able to break down and get rid of the antidepressants so risk is very small
- Sertraline & paroxetine only get into the breast milk in very small amounts hence recommended.

POST STROKE DEPRESSION

- Occurs in 30-60%
- More cognitive dysfunction, fatigue and anxiety
- 5-10% have emotional lability
- Does not improve spontaneously
- Fluoxetine is recommended

DEPRESSION AND PHYSICAL HEALTH PROBLEMS

Sertraline has the most evidence in heart disease. Both sertraline and citalopram are suggested for patients with co-morbid physical health problems. Fluoxetine and paroxetine are to avoided due to the higher risk of interaction with other medications.

ECT- EXPLANATION

What is ECT?

ECT stands for electoconvulsive therapy. Most people who have ECT are suffering from severe depression not responding to other medications.

Is it the same as shock treatment?/Isn't it barbaric to still give it in this day and age?

In the past it used to be known as shock treatment. ECT involves passing a small amount of electrical current through the brain to produce a fit. The idea developed from the observation that, in the days before when there was no effective medication, some people with depression who also had epilepsy, seemed to feel better after having a fit. The effect is thought to be due to the fit rather than the electrical current. The circumstances under which it is given now, is much safer than previously.

How does it work?

One of the causes of depression is the chemical imbalance in the brain. The fit alters this imbalance.

What is the procedure?

- Consent: Explain that this is a treatment that we are considering and he can withdraw his consent at any time, even after the first treatment. Reassure him that you are only there to explain to him about ECT and answer his questions.

- Blood tests, chest X-ray, ECG: to make sure it is safe to have general anaesthesia and to check his physical health

- Must not eat or drink anything for 6 hours before ECT

- You will be brought to the ECT suite with an experienced nurse with whom you are familiar so that things can be explained to you as they happen. There will be an anaesthetist and also your psychiatrist.

- You will be connected up to monitoring equipment to check your heart rate, blood pressure, oxygen levels, and an EEG machine that will monitor brain waves.

- The anaethetist will then give you an injection of an anaesthetic medication and once you are asleep, he will also give you a muscle relaxant to prevent any injuries while you have a fit. The anaesthetist will give you oxygen to breathe as you fall asleep.

- You will then be given the ECT treatment. The fit will last for 20-50 seconds.

- Once completed and you are gradually waking up you will be taken to the recovery area where a nurse will monitor you (and do some checks like your blood pressure and oxygen levels) until you are fully awake. You may feel disorientated when you initially wake up and feel sick. This will wear off after about 30 mins.

How often will this be given?

- It is usually given twice a week.

- It is impossible to predicy how many treatment you might require. Usually within 4-5 treatments there will be noticeable improvement.

- Normally the course lasts for 6-8 treatments, although sometimes 12 treatments may be needed.

- After 12 treatments if there is no response, then it is unlikely you will respond to ECT.

What are the side effects?

- Short term: headache, muscle ache, nausea, confusion- these usually settle quickly; There will be some temporary loss of memory for the time immediately before and after the treatment. Due to use of general anaesthesia, there is 1:50,000 risk of death.

- Longer term: Long term memory impairment occurs in about 1 in 10. This can be for past events which could be significant to you like birth of your child. Most people find these memories return when the course of ECT has finished and a few weeks have passed. However, some complain that there is a permanent loss of memory. We will continue to assess your memory as you go along and if it gets worse we can reduce the frequency of sessions to once a week or give ECT on side of the brain. (Unilateral- fewer side effects but less effective)

How well does it work?

8 out of 10 people respond well. They feel more optimistic, feel like themselves (and less suicidal)

What will happen if I don't get ECT?

You might take longer to recover, become physically ill secondary to the mental illness or even suicidal.

Any other alternatives to ECT?

ECT is one of the treatments that we propose for your depression. If you do not agree to ECT there are alternatives including medication change, a new medication, intensive CBT. Although this will all have most likely been tried already.

Will you force this on me?

No. I am here to discuss the option of ECT with you. As mentioned earlier, we will not do anything without consent. However, sometimes people become so unwell they are unable to understand all of the issues and consent or give proper agreement to the treatment. ECT can then be given under Mental Health Act.

CHAPTER 7   ANXIETY AND SOMATOFORM DISORDERS

ANXIETY SYMPTOMS

Autonomic arousal symptoms:

- Palpitations/pounding of heart
- Sweating
- Trembling/shakes
- Dry mouth

Chest and abdomen:

- Difficulty breathing
- Feeling choking
- Chest pain
- Nausea

Mental state:

- Feeling dizzy, unsteady, light headed
- Derealization, depersonalization
- Fear of losing control, "going crazy"
- Fear of dying

General symptoms:

- Hot flushes/cold chills
- Numbness/tingling sensations

- Muscle tension, aches and pains

- Restlessness, inability to relax

- Feeling on edge

- Difficulty in swallowing/sensation of lump in the throat

Other non specific symptoms:

- Easily startled

- Poor concentration

- Persistent irritability

- Decreased sleep

Anticipatory anxiety

AGORAPHOBIA

a) Psychological/autonomic symptoms must be primarily manifestations of anxiety

b) Anxiety must be restricted to at least 2 of the following situations: crowds, public places, travelling away from home and travelling alone

c) Avoidance of phobic situation must be a prominent feature

- Associated with significant emotional distress

- Recognises that the thoughts are excessive or unreasonable

- Thought of collapsing and being left helpless in public (lack of immediately available exit)

SOCIAL PHOBIA

a) Fear of scrutiny by other people leading to avoidance of social situations

b) Usually associated with low self-esteem and fear of criticism

c) May present as a complaint of blushing, hand tremor, nausea or urgency of micturition

d) Symptoms may progress to panic attacks

PANIC DISORDER

a) The essential feature is recurrent attacks of severe anxiety (panic), which are not restricted to any particular situation or set of circumstances and are therefore unpredictable

b) Dominant symptoms include sudden onset of palpitations, chest pain, choking sensations, dizziness, and feelings of unreality (depersonalization or derealization)

c) There is often also a secondary fear of dying, losing control, or going mad

GENERALISED ANXIETY DISORDER

a) Anxiety is generalized and persistent but not restricted to, or even strongly predominating in, any particular environmental circumstances

b) The dominant symptoms include complaints of persistent nervousness, trembling, muscular tensions, sweating, lightheadedness, palpitations, dizziness, and epigastric discomfort

c) Fears that the patient or a relative will shortly become ill or have an accident are often expressed

ANXIOUS (AVOIDANT) PERSONALITY DISORDER

a) Persistent feeling of apprehension/fear

b) Belief that one is socially inept

c) Excess belief of being criticized/rejected in social situations

d) Unwillingness to become involved with people unless certain of being liked

e) Restrictions in lifestyle because of need for physical security

f) Avoidance of activities that involve significant interspersonal contact because of fear of rejection

DIFFERENTIAL DIAGNOSIS FOR ANXIETY SYMPTOMS

- Organic cause (cushings, hyperthyroidism, phaeochromocytoma, arrhythmias, COPD, asthma, MI epilepsy, vestibular dysfunction)
- Psychosis
- Mood disorder
- Somatoform
- Substance misuse

SOCIAL PHOBIA- EXPLANATION

What is social phobia?

Phobia is a strong fear of a thing or event which is out of proportion to the reality of the situation. In social phobia you get anxious when you are with other people because you worry that they may be critical of you or that you may do something embarrassing. This can be so severe that you can't bear to be to speak or be with other people and avoid social situations altogether. About 5 in 100 people have some level of social phobia, 2-3 times more common in women.

Symptoms

As mentioned previously

Causes?

- Those who have particularly high standards for their behaviour in public
- Stammering as a child
- Being stuck at the normal stage of shyness that all children go through between 3 – 7 years old
- Certain thoughts tend to kick in when you enter a social situation and will make you anxious (thinking errors)
- Certain safety behaviours like use of alcohol, avoiding eye contact, avoiding saying anything personal about yourself reinforce the phobia

- Thinking over and over about a social situation, before or after, tends to make you focus on past 'failures' and makes you more self critical

Comorbidity

Depression, drug and alcohol misuse, agoraphobia

Self-help?

- Self-confidence or assertiveness course.
- Relaxation techniques
- Write down the thoughts and images you have about yourself so that you can start to change them
- Listen more to what people say
- Start to stop using your 'safety behaviours', beginning with the easiest.
- Break down a worrying situation into a number of steps

Psychological treatments

- Social skills training: This teaches simple social skills like how to talk to a stranger and helps you to feel more comfortable in social situations

- Graded self-exposure: You make a list of all the situations that you find frightening, and then put them in order, from the least frightening to the most frightening. You start with the least frightening situation and, with the support of your therapist, keep yourself there until you stop feeling anxious. You then move on to the next one. You get through the situations one by one from the least anxiety provoking to the most.

- CBT: Social phobia is strongly related to yours thoughts about yourself, the world and people around you. CBT helps you to change the way that you think about yourself and others.

Medications

- Antidepressants: SSRIs or MAOIs. Usually start to work within 6 weeks, but can take up to 12 weeks to have full effect. 50% relapse when antidepressants are stopped.

- Beta-blockers: In low dose they can be used to control the physical symptoms of social phobia and can be taken shortly before meeting people or before speaking in public.

PANIC ATTACKS- EXPLANATION

Anxiety is the normal human feeling of fear that we all experience when faced with threatening or difficult situations. If the feelings of anxiety are too strong it can affect our day to day lives.

When we are exposed to things that we fear, we either face them or try to run away from them (fight/flight response). When this happens, we become more alert, our heart beat speeds up, muscles get tense, we breathe faster to get more oxygen to our muscles, consequently we breathe out carbon dioxide which produces strange sensations like dizziness, tingling in hands & feet, chest pains, breathlessness. When this happens we breathe faster and this makes our symptoms worse.

Normally we use the muscles in the tummy to breathe but when we breathe faster it is teh chest muscles that are used more and hence we get chest pain, which can be misinterpreted as a heart attack. Panic attacks do not cause heart attacks. However, if you repeatedly get chest pain it is advisable that you see a doctor to rule out any underlying heart disease.

Two methods to reduce the symptoms are breathing into a paper bag which will increase the level of carbon dioxide in your body and by learning controlled breathing. You will be able to manage your panic attacks if you recognize the symptoms are harmless and to recognize how and when they happen.

124

Remember to use slow breathing exercise. Take a deep breath in and breathe out slowly saying the world relax to themselves in a calm soothing manner every time you breathe out. It should subside in a minute or two and hopefully avoid a panic attack.

SYSTEMATIC DESENSITIZATION/GRADED EXPOSURE WITH RELAXATION

- The therapist will teach you relaxation exercises to control to reduce anxiety.
- You then make a list of situations which you find difficult to face from least difficult to most difficult. Then start by facing the easiest situation and moving onto next situation.
- Practice steps until there is no longer a cause for anxiety.
- Need to practice regularly and gradually.

PANIC DISORDER- MANAGEMENT (NICE GUIDANCE)

- Treat avoidance by exposure
- Psychoeducation, managing thinking errors
- No benzodiazepines, sedating antihistamines or antipsychotics
- CBT weekly- 7-14 hours in total
- SSRI (Paroxetine, fluoxetine, citalopram, setraline)- if no improvement after 12 weeks, imipramine or clomipramine; continue for 6/12

- Bibliotherapy (written material to help people understand their psychological problem and learn ways to overcome them based on CBT principles)

GENERALISED ANXIETY DISORDER- TREATMENT

- CBT weekly 16-20 hours total

- Short term benzodiazepines (for 2-4 weeks only)

- SSRI- paroxetine is licenced; continue for 6/12

- If no improvement try venlafaxine

OBSESSIVE COMPULSIVE DISORDER

For a definitive diagnosis , obsessional symptoms or compulsive acts, or both, must be present on most days for at least 2 successive weeks and be a source of distress or interference with activities. The obsessional symptoms should have the following characteristics:

- Recognised as the individual's own thoughts or impulses

- At least one thought or act that is still resisted unsuccessfully

- No pleasure from thoughts

- Unpleasantly repetitive

OCD- HISTORY

Obsessions:

- Thoughts, ideas, images, ruminations, doubts (including aggressive and religious thoughts)
- Own thoughts
- Repetitive and intrusive
- Excessive and unreasonable
- Resistance unsuccessful
- No pleasure

Anxiety symptoms after getting obsessive thoughts

Compulsions:

- Do you try to neutralize these thoughts?
- Counting, washing, checking, rituals, hoarding
- Avoidance

Symptom severity:

- Do you find it difficult to make decisions even for simple trivial things?
- How hard do you try and resist?
- How much control do you have over them?
- Effect on work, relationship and family

Co-morbidity:

- Schizophrenia, depression, anxiety, panic attacks
- Body dismorphic disorder
- Substance misuse

- Tourette's (sudden sounds or movements you cant control?),

Premorbid personality:

- Anankastic personality traits (do you tend to keep things in an orderly way?)

Insight

Risk assessment

OCD TREATMENT

Brief CBT including ERP using self help materials or by phone or group CBT

Review at 12 weeks

Adults mild impairment offer more intensive CBT or SSRI (for OCD fluoxetine, paroxetine, citalopram, fluvoxamine, sertraline for BDD fluoxetine) if can't do self help.

Moderate impairment CBT or SSRI

Consider clomimpramine – potentially dangerous side effects – convulsions (2% - 0.5-1% on SSRIs) and cardiotoxicity, sexual dysfunction in 80% (30% for SSRIs)

Augmentation – clomiprmaine and citalopram, add antipsychotic to SSRI or clomipramine, additional CBT

Severe – CBT(with ERP) & SSRI

BDD augmentation – additional CBT, add buspirone to SSRI

Continue for 12 months

Admit if

Risk to life

Severe self neglect

Extreme distress/functional impairment

Adequate treatment trials not successful

Co morbidities

Cant get to clinic due to OCD

OCD- EXPLANATION

What is OCD?

Obsessive compulsive disorder is a mental illness which affects 1 in 50 people, in which you get repeated thoughts that come into your mind that you can't get rid of (obsessions). They keep coming to your mind even when you try to resist them.

Obsessions can be:

- Thoughts

- Pictures in your mind

- Doubts (example- that you have caused an accident)

- Ruminations (example- endlessly argue with yourself over simple decisions)

You may feel anxious because of these and therefore carry out your ritual or compulsive behaviour

Compulsions include:

Correcting obsessional thoughts- by thinking correcting thoughts

Rituals- such as washing your hands, checking

Avoidance- of any reminder of the worrying thoughts

Hoarding

Causes

There is no single cause

- It tends to **run in the family**

- **Stressful life events**

- **Chemical imbalance** of serotonin

- **Personality**- if youre neat, meticulous person, sometimes when this goes too far, it can develop into OCD

- **Ways of thinking**- if you have particularly high standards of morality and responsibility you may feel especially bad when you have these thoughts. So you start to watch out for them which makes them more likely to happen.
- **In children it can start after a bacterial infection**

Treatment

Psychological therapies

- Cognitive therapy: CBT (helps to change your reaction to the thoughts instead of trying to get rid of them)
- Exposure and response prevention: is based on exposing yourself to the thoughts or situations that cause anxiety. If you stay in a stressful situations for long enough the anxiety goes away. . Put the situations or thoughts you fear the least at the bottom and work up. This is best done is small steps firstly making a list of thoughts or situations you fear. Needs to be repeated daily for 1 to 2 weeks. You will see how your anxiety rises, then falls. Don't move to the next stage until you have overcome the last.

Medications

- Drug treatment includes use of antidepressant (SSRI)
- Combination antidepressants like clomipramine plus citalopram
- Treating comorbidities like including anxiety, depression
- Adding antipsychotic medication

SELF HELP

- Expose yourself to the bad thoughts- the more regularly you do this the better you can control them. You could record them on paper or audio and listen back to them while at the same time resisting the compulsive behaviours.

- Resist the compulsions but not the obsessional thoughts

- Don't use alcohol to control your anxiety.

- Self help books might also be helpful

- Support groups like OCD UK

SUGGESTIONS FOR CARERS

- OCD sufferers' behaviour can be quite frustrating. Try not to get angry with them as they are trying their best to cope

- Educate yourself about OCD.

- It may take a while for someone to accept that they need help. Encourage them to read about OCD and talk it over with a professional.

- Help with exposure treatments by reacting differently to the compulsions:
  - encourage tackling anxiety provoking situations
  - don't take part in rituals or checking

- don't reassure them that things are alright.

Prognosis

3 out of 4 improve with exposure and response prevention, but 1 in 4 will get symptoms again

6 out of 10 improve on antidepressants. Half of those who stop medication will get unwell again in the months afterwards.

BODY DYSMORPHIC DISORDER- HISTORY

- Part/ aspect of the body affected?

- Any other parts affected?

- Reasons for preoccupation of the deformity

- Strength of belief (differentiate between delusion, obsession and overvalued idea)

- Previous investigations for the defect (including surgeries)

- Behaviours:
    - Avoidance of social situations
    - Repeated checking of the perceived defect
    - Mirror gazing
    - Elaborate grooming rituals/attempting to camouflage the defect
    - Reassurance seeking

- Impact:

- Work

- Relationships

- Meeting new people (staring/ talking/ laughing)

- Exclude: OCD, depression, social phobia, psychosis, substance misuse, delusional disorder, anxiety/panic disorder, somatisation disorder

- Risk assessment:

  - Suicidal ideation

  - Thoughts of actually performing the surgery themselves

  - Persistent refusal to accept the advice

- Past psychiatric history

- Family history of affective disorders/OCD

ACUTE STRESS REACTION

- Immediate and clear temporal connection between the impact of an exceptional stressor

- Onset usually within a few minutes

- Mixed and changing picture- initial state of "daze", depression, anxiety, anger, despair, overactivity and withdrawal (no type of symptom predominates for long)

- Resolve rapidly usually diminishing by 24-48 hours and are minimal after about 3 days

POST TRAUMATIC STRESS DISORDER

- Symptoms arose within 6 months of a traumatic event of exceptional severity
- History of the traumatic event
- Repeted intrusive recollection or reenactment of the event:
    - Images, thoughts, flashbacks, dreams
    - Acting or feelinga s if the event was reoccuring
    - Distressing dreams of the event
    - Intense psychological distress at exposure to cues that represent the traumatic event (e.g. when talking about it)
- Feeling of detachment from others- depersonalization/derealization
- Emotional numbing
- Persistent avoidance:
    - Efforts to avoid thoughts, feelings, conversations associated with the trauma
    - Efforts to avoid activities, places or people that arouse recollections
- Increased arousal:
    - Difficulty falling/staying asleep
    - Irritability or outbursts of anger

- Poor concentration

- Hypervigilance (always on edge)

- Exaggerated startle response

- Physical symptoms:

    - Aches and pains

    - Headaches

    - Feelings of panic and fear

- Comorbidity: Affective disorder, substance misuse, anxiety disorders

- Management

    - Don't offer debriefing

    - Mild symptoms for less than 4 weeks- watchful waiting, follow up within 1 month

    - Severe PTSD in the first month- trauma focused CBT (8-12 sessions) or EMDR

    - First line- Paroxetine or Mirtazapine

    - Second line- Amitriptyline or Phenelzine

    - Continue the mediactions for at least 12 months

    - If no response use adjunctive Olanzapine

    - Treat co-morbidities accordingly

PTSD- EXPLANATION

What is PTSD?

- PTSD can start after any traumatic event when we are in danger or our life is threatened eg serious road accident, terrorist attack, being diagnosed with a life threatening illness, unexpected injury or violent death of someone close.

- Symptoms are worse if the event is sudden, especially violent, man made or involves many deaths.

- It can also occur from less severe long lasting traumas like ongoing physical abuse at home.

- Most people will develop post traumatic stress for 6 weeks or so. Symptoms usually develop within 6 months of the event happening.

- 1 in 4 exposed to potentially traumatic event develop PTSD.

- Can affect any age group.

Symptoms

As mentioned above

Causes

- Remembering things clearly after a shock helps to understand what happened and help you to survive.

- Flashbacks make you think about what has happened and therefore decide what to do if it happens again

- Avoidance and numbing help you to stop becoming tired and distressed from remembering a trauma and reduces the number of times the event is replayed in your mind.

- Being 'on guard' means that you can react quickly if another crisis happens and give you and adrenaline rush to keep going after the event.

- Vivid memories tend to keep your adrenaline levels high meaning you are tense irritable and unable to sleep.

- In the brain, a part of the brain called hippocampus processes memories. In PTSD, high levels of stress hormones, like adrenaline, may stop it from processing the memories of the event, leading to ongoing flashbacks and nightmares.

What can I do to help myself?

- Try to get back to usual routine

- Talk to someone you trust about what happened

- Do go back to where the event happened

 Relaxation exercises

- Balanced life style- healthy diet and regular exercise

- Be careful about driving as your concentration will be poor and accidents are more likely

- Don't smoke or drink lots of alcohol or caffeine

Treatment

- Talking therapies like CBT focus on the trauma. You cannot change what has happened, but you can change the way you think about itIt helps to think differently about your traumatic experience.

- EMDR (Eye movement desentisation and reprocessing)- technique which uses eye movements to help the brain process flashbacks and make sense of the traumatic experience

- Antidepressants

- Talking to people in a group who have had similar experiences might help

- Body focussed therapies physiotherapy, osteopathy, massage, acupuncture, reflexology, yoga, meditation can help to control distress and reduce the feeling of being 'on guard'

Prognosis

- CBT, EMDR and medication are all effective

- 50% of patients do improve fully

ADJUSTMENT DISORDERS

- State of subjective distress and emotional disturbance, usually interfering with social functioning and performance, arising in the period of adaptation to a significant life change or a stressful life event

- Onset within one month of life change/stressor

- Symptoms do not usually last longer than 6 months

GRIEF REACTION

- Terminology:

    - Bereavement describes the situation of having lost someone significant

        to them

    - Mourning is the process of adjusting to bereavement

    - Grief is the personal experience of mourning

- Normal grief:

    - Lasts up to 12 months (average 6 months)

    - Stages- shock, denial, anger, bargaining, depression and acceptance

- Shock:

    - Lasts for hours to days (average of 2 weeks)

    - Intense mental and somatic distress

    - Numbing, denial and disbelief

    - Appears dazed or immobile

    - Anxiety symptoms

- Preoccupation with the deceased (weeks to 6 months)

    - Searching behaviour in the hope of reunion

    - Spending long periods thinking about them

    - Reliving memories

    - Anger directed at third parties

- Dreaming frequently about the deceased

- Brief hallucinatory experiences (auditory and visual)

- Social withdrawal

- Fluctuating depressive symptoms

- Resolution (weeks to months)

    - Subjective feeling of acceptance

    - Reorganization of life

- Abnormal grief (prolonged, intense, delayed or absent):

    - Associated with sudden death, very close, dependent or ambivalent relationship with the deceased, history of insecurity or psychiatric illness

    - Preoccupation with feelings of worthlessness (thoughts of death)

    - Excessive guilt

    - Marked slowing of thoughts and movements

    - Prolonged period of not being able to function properly

    - Hallucinatory experiences other than voice or image of deceased

    - Mummification (still place setting for them at dinner or not moving the personal things of the deceased)

- Beware of depression in grief when there is:

    - Psychomotor retardation

    - Generalized guilt and suicidal thoughts after the first month

    - Previous history of depression

- Intense grief or depressive symptoms early in the grief reaction

- Few social supports

- Co-morbidity: Substance misuse

- Assess social history & functioning

- Treatment

    - Sleepless nights- short course of sleeping tablets

    - Voluntary/religious organizations like CRUSE

    - Depression- antidepressants

- Help from carers:

    - Spend time with the bereaved person and give them time to grieve

    - Let them, if they want to, cry with you and talk about their feelings

    - Don't tell them to pull themselves together

    - Help out with practical things like shopping, cleaning etc

    - Try to be around at particularly painful times, such as anniversaries

DISSOCIATIVE DISORDERS

- Partial or complete loss of the normal integration between memories of the past, awareness of identity and immediate sensations, and control of bodily movements

- Tend to remit after a few weeks or months, particularly if their onset is associated with a traumatic life event

- Onset is usually associated with traumatic events, insoluble problems and interpersonal difficulties
- No evidence of physical disorder
- There is evidence that the loss of function is an expression of emotional conflicts or needs
- The symptoms may develop in close relationship to psychological stress, and often appear suddenly

DISSOCIATIVE AMNESIA

- Loss of memory, usually of recent important events (too extensive to be explained by ordinary forgetfulness or fatigue)
- Memory loss is usually partial and selective
- Perplexity, distress, and varying degrees of attention seeking behaviour
- Calm acceptance of memory loss
- Purposeless local wandering may occur
- Self neglect may be evident
- Rarely lasts more than a day or two

DISSOCIATIVE FUGUE

- Features of dissociative amnesia
- Purposeful travel beyond the usual everyday range

- Maintainance of basic self care

- Simple social interaction with strangers

DISSOCIATIVE MOTOR DISORDER

- Complete or partial loss of ability to perform voluntary movement

- Various degrees of incoordination, ataxia, inability to stand

DISSOCIATIVE ANAESTHESIA AND SENSORY LOSS

- Partial or complete loss of some sensation over part of body (light touch, vibration, etc.)

- Partial or complete loss of hearing, smell, vision

- Visual disturbances like loss of acuity, general blurring of vision or "tunnel vision" (loss of vision is rarely total)

- General mobility and motor performance are often well preserved

SOMATIZATION DISORDER

- At least 2 years of multiple and variable physical symptoms for which no adequte physical explanation has been found

- Persistent refusal to accept the advice or reassurance of several doctors that there is no physical explanation for the symptoms

- Asks for treatment to remove the symptom

- Some degree of impairment of social and family functioning

- Rule out physical disorder, depression, anxiety, hypochondriacal disorder and delusional disorder

- History taking (in addition to the above):

  - History of excessive use of medical services and alternative therapies

  - Diagnostic procedures and operations

  - Ask the patient- if the investigation is performed and the results come back normal what will you think?

  - Link the physical symptoms with psychological issues- for example, "You have told me you have got back and you have also told me you are under a lot of stress at work. Do you think they are linked?"

- Management

  - Explain empathically that this is a functional illness

  - Empirical use of potentially beneficial treatment like antidepressants even if depressive features not present

  - Physiotherapy to aid regaining functional loss

  - Psychotherapy

  - Pain management

HYPOCHONDRIACHAL DISORDER

- Belief in at least one serious physical symptoms despite evidence to contrary (for at least 6months)

- Persistent refusal to accept advice and reassurance of doctors

- Requests investigations to determine nature of the illness
- Can have persistent preoccupation with deformity/disfigurement (body dysmorphic disorder)
- Rule out somatization disorder, depressive disorders, delusional disorder, anxiety and panic disorder

NEURASTHENIA (CHRONIC FATIGUE SYNDROME)

- Persistent and distressing complaints of increased fatigue after mental effort or bodily weakness and exhaustion after minimal effort
- At least 2 of - feelings of muscular aches and pains, dizziness, tension headaches, sleep disturbance, inability to relax, irritability and dyspepsia
- Any autonomic or depressive symptoms present are not sufficiently persistent and severe to fulfil the criteria for any of the more specific disorders
- Rule out mood, generalised anxiety disorder, post concussional syndrome, postencephalitic syndrome
- Painful lymph nodes without pathological enlargement, sore throat, palpitations in the absence of a cardiac cause
- Cognitive dysfunction- difficulty thinking, decreased concentration, impairment of short term memory, difficulties in word finding and planning
- Flu-like symptoms

- Investigations:
  - FBC, U& Es, LFTs, TFTs, ESR, CRP, glucose, calcium, CK, screening for gluten sensitivity, (serum ferritin- for children and young people only)
  - Urinalysis- protein blood, glucose
- Management:
  - Symptomatic management (example- manage nausea conventionally with advice on eating little and often, snacking on dry starchy foods and sipping fluids; antiemetic if severe)
  - Exclusion diet not recommended
  - Sleep management
  - Rest periods- limit to 30 mins at a time, introduce low level physical and cognitive activities
  - Relaxation
  - Diet – well balanced, eat regularly
  - Education and employment
  - Relapse prevention strategies
  - CBT and/or graded exercise therapy(GET)
  - Activity management- goal oriented, person centred, splitting the activities (if more severe may need to do a tailored activity plan at home, keeping a diary and gradually building up)
  - Complimentary therapies are not recommended
- Prognosis:

- 6 out of 10 people feel better with either CBT or Graded Exercise Therapy, although quite a few people with CFS have reported that GET tends to make them more tired, not less.

- About 1 in 4 treated patients rate themselves as completely recovered from their CFS/ME after CBT

CHAPTER 8   **EATING DISORDERS**

ANOREXIA NERVOSA

- Body weight maintained at 15% below that expected, or BMI is 17.5 or less

- Weight loss is self induced by avoidance of fattening foods

- Self induced vomiting, self induced purging, excessive exercise, use of appetite suppressants and/ or diuretics

- Body image distortion- dread of fatness persists as an intrusive, overvalued ideas and the patient imposes a low weight threshold on themselves

- Widespread endocrine disorder- amenorrhoea in women and loss of sexual interest and potency in men

EATING DISORDER- HISTORY

- Eating pattern:

    - Start of symptoms

    - Changes over time

    - Routine pattern (typical day- quantity and frequency)

    - Counting calories

    - Premorbid obesity

    - Reasons for avoiding food, especially fatty food

- Behaviours:

- Bingeing (irresistible craving for food, episodes of large amounts of overeating in short periods)

- Purging (to get rid of the calories that you have eaten)

- Self induced vomiting

- Use of laxatives/thyroxine/diuretics/amphetamines/emetics

- Excessive exercise

- Psychological issues:

  - Current height and weight (including highest and lowest weights)

  - Ideal weight

  - Ambition and role model

  - How much weight loss over how long

  - Has anyone mentioned you are fat?

  - How do you feel when you look at yourself in the mirror?

- Physical symptoms:

  - Tiredness

  - Dizzy spells

  - Constipation

  - Muscle cramps

  - Thyroid problems

  - Menstrual changes

  - Decreased libido

- Medical problems: Chronic debilitating diseases, intestinal disorders, tumours

- Explore issues at home (enmeshed family structure):
    - How would you describe your family? (too intrusive/ controlling?)
    - Whom are you closest to in the family?
    - Would you like to be more independent?

Co-morbidity: Depression, anxiety, personality disorder, substance misuse

Rule out OCD and psychosis (Is the food poisoned?)

Family history of eating disorder

Relationship history

Social history- social activities, social pressure

Premorbid personality- coping with stress, self esteem issues, lack of control, maturity fears

ANOREXIA NERVOSA- EXPLANATION

What is wrong with her/him?

We all have different eating habits. Sometimes when people are under stress, they find it easy to focus on their eating habit, instead of things that are bothering with them. If this goes on for long enough, that can develop in to an eating disorder. In anorexia nervosa, people keep their body weight low by dieting, vomiting or exercing excessively. They have an intense fear of becoming fat, hence avoid food.

When does it start?

- It begins usually in the teenage years

- It is 10 times more common in girls

- 1 in 150 in 15 year old girls and 1 in 1000 in 15 year old boys

Causes

There is no simple answer to this.

- **Social pressure**: Social surroundings powerfully influence our behaviour. Currently "thin is beautiful" and there is a lot of focus on size zero image. So, at some time or the other, most of us try to diet and sometimes, people diet too much and slip in to anorexia.

- **Lack of an "off switch":** Most of us can only diet so much before our body tells us that it is time to start eating again. Some people lack this ability.

- **Genetics**: Can run in family

- **Low self esteem** issues: People with anorexia nervosa see their weight loss as apositive achievemnt that can help increase their confidence and self esteem, thus giving them a feeling of control over their life.

- **Puberty**: Anorexia can reverse some of the physical changes, particularly the sexual ones. This may help to put off the demands of getting older.

- **Family**: Overprotective family, enmeshment, rigidity; saying " no" to food may be the only way to express feelings, or any say in family affairs.

- **Disturbed body image**

152

- **Life events**: Anorexia nervosa has been related to life events like sexual abuse, physical illness, upsetting events like death or break ups, or even marraiges and leaving home.

Physical symptoms

- Starvation:
    - Harder to eat because stomach has shrunk
    - Constipation and tiredness
    - Affects growth- brittle bones leading to fractures
    - Affects ability to become pregnant.
    - Death
- Vomiting:
    - Lose the enamel on teeth
    - Puffy face (swelling of the salivary glands)
    - Irregular beating of the heart (vomiting disturbs the balance of salts in the blood)
    - Tiredness
    - Damage to kidneys
    - Fits
    - Unable to get pregnant
- Regular laxative use:
    - Persistent stomach pain

-   Swollen fingers

-   Constipation

-   Swinging body weight

SELF HELP

-   Try to stick to regular mealtimes

-   Keep a diary of what you eat and your thoughts and feelings

-   Be honest with yourself

-   Join a self help group/contact eating disorders association

-   Make sure you know what a reasonable weight is and why

-   Remind yourself you don't always have to be achieving things

-   Remember that the more weight you lose the more anxious and depressed you will feel

Treatment

-   Inpatient or outpatient treatment depends on the physical and mental state of the patient

-   Manage physical health- bloods (FBC, U & Es, LFTs, TFTs, hormonal assay (cortisol, GH, LHRH, LH, FSH, oestrogens and progestogens), cholesterol, amylase, ESR, glucose), ECG (ECHO if required)

-   Regular weight checks

-   Criteria to admit

- BMI <13.5

- Syncope

- Proximal myopathy

- Hypoglycaemia

- Electrolyte imbalance

- Petechial rash/platelet suppression

- Risk of suicide

- Chronic (>5 years)

- Co-morbidity with impulsive behaviour

- Intolerable family situation

- Extreme social isolation

- Failure of outpatient treatment

- Inpatient management:

  - Restore and monitor weight- 0.5-1kg per week (outpatient 0.5kg per week)

  - Initially (first 3-7 days) soft diet of approximately 30-40kcals/kg/day in small portions throughout the day

  - Slowly increase caloric intake 200-300 kcal every 3-5 weeks

  - Oral replacement to prevent dehydration

  - BMI needs to be charted

  - Bone scan for osteoporosis- may need calcium 1500mg/day and Vitamin D 400 IU per day

- Correction of serum electrolytes

- Beware of risks of refeeding (hypocalcaemia, hypophosphataemia, hypokalemia, hypomagnesemia, cardiac decompensation)

- Nursing- monitor suicidaility, impulsive behavior, self harm, support at meal times

- Psychological treatment

- Psychological treatment: Focus on eating habits, attitude towards food, weight and shape, and the deelings in relation to gaining weight (If less than 18 Family Therapy better than CBT)

    - CBT (establishes links between thoughts, feelings, behaviour and their symptoms)

    - Family therapy (Involves family taking part in sessions. Concentrates on the eating disorder and how it affects family relationships and how these may have led to the development of the eating disorder. Sessions may focus on how the family can work together to support each other and work through difficult issues they are facing)

    - CAT (The therapist works with the patient to help them achieve positive changes in their lives and work towards the future)

    - Focal psychodynamic (Aims to identify and focus on difficulties that may have occured in the early stages of life and which are responsible for the current problems)

- IPT (Helps to identify and address current interpersonal problems, like low self esteem)
- Problem solving and motivation interviewing
- Other therapies: art therapy, drama therapy
- Medications: to treat co-morbidities like OCD, depression

Poor prognostic factors

- Severe weight loss
- Long duration of illness
- High level of reversing behaviours (exercise, vomiting)
- Failure to attain normal weight
- Additional co-morbidity
- Poor premorbid psychological adjustment

Prognosis

- More than half of the sufferers make a recovery
- Full recovery can happen even after 20 years of diagnosis
- Severely ill- 1 in 5 will die

BULIMIA NERVOSA

- Persistent preoccupation with eating

- Irresistable craving for food

- Episodes of large amounts of over eating in short periods

- Counteract the fattening effects by self induced vomiting, purgative abuse, alternate periods of starvation use of drugs like laxatives

- Psychopathology- morbid dread of fatness

- Sets sharply defined weight threshold

- History of anorexia nervosa

BULIMIA NERVOSA- EXPLANANTION

- Start with general introduction of eating disorder (as mentioned previously)

- Is an illness in which people feel they have lost control over their eating. They are caught in a cycle of eating large quantities of food (called binge eating) and then vomiting afterward. They have a fear of fatness. They often use laxatives and diuretics or exercise excessively and starve in order to prevent gaining weight.

- Affects 1-3% of adolescent females; starts mid teens

- Causes:
  - Can run in families
  - Environment (rigid dieting)

- Chemical imbalance of serotonin

- Personality issues

- Treatment:

  - Usually as an outpatient, unless suicidal, physical problems, pregnant or refractory

  - High dose SSRI- Fluoxetine 60 mgs OD for 1 year

  - CBT (16 -20 sessions over 4-5 months)

  - Interpersonal therapy

  - Treat any comorbidity

  - Regular dental reviews for those who vomit and dental hygiene advice (no brushing after vomit, non acid mouthwash, limit acidic foods)

- Poor prognostic factors:

  - Depression

  - Personality disturbance

  - Greater severity of symptoms

  - Longer duration of symptoms

  - Low self esteem

  - Substance abuse

- Prognosis: 50% recover, 40% have residual symptoms; recovery usually takes place slowly over a few months or many years

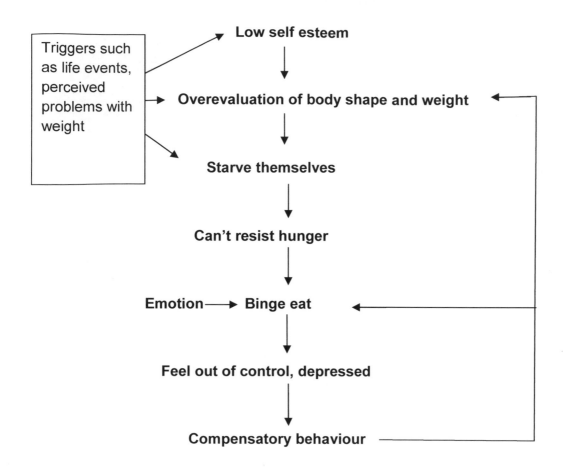

CHAPTER 9   PUERPERAL DISORDERS

POST NATAL DEPRESSION & PUERPERAL PSYCHOSIS- HISTORY

- Risk factors:

    - Planned pregnancy or not

    - Problems during pregnancy/labour

    - Partner- support during pregnancy/ after birth

- Relationship with the baby

    - How do you feel

    - Coping

    - Does he sleep well

    - Does he cry too much

    - Have you been losing your temper with the baby

    - Anxiety about well being of the baby

    - Does he have any problems

    - Worried that someone might take him away

- Affective symptoms

- Psychotic symptoms

- MSE including:

    - How do you feel in yourself?

    - How do you feel as a mother (useless/ worthless)?

- Do you blame yourself for something you have done or thought?

- Risk assessment

    - Self

    - Baby

- Past psychiatric history

- Family history

- Social supports

PSEUDOCYESIS

- Symptoms of pregnancy

    - Amenorrhoea

    - Tender breasts

    - Weight gain

    - Morning sickness

    - Abdominal distension

    - Feeling baby move

- Attitude to pregnancy

- Past medical history

- Past gynaecological history

    - Pregnancies

- Miscarriages

- Sterilisation

- Past psychiatric history

- Relationship history

- Social history

- Alcohol and drug history

- Risk assessment

- Treatment:

  - Support and insight oriented psychotherapy

  - Trial of SSRIs

POSTNATAL DEPRESSION- EXPLANATION

What is post natal depression?

- Is becoming depressed after having a baby

- Around half of mothers become weepy and tearful on the 3rd or 4th day (baby blues). This passes after a few days.

- Postnatal depression is quite common and affects 1 in 10 women.

- It usually starts within a month of delivery but can start up to 6/12 after

Symptoms

- Depressive symptoms

- Difficulty bonding with the baby or unable to cope with the baby

- Anxious about the baby, despite strong feelings of love for the baby

Causes

There is no single cause.

- Previous history of depression

- Lack of supportive partner/family

- Stressful life events like bereavement, money worries, unemployment can precipitate

- Premature/ sick baby- constant worry

- Loss of own mother when a child

- Sometimes it can start without any obvious reason

Do they harm their baby?

- They usually don't harm the baby but worry that they might.

- Some parents at times may feel like hitting or shaking the baby, not just those with post natal depression

Treatments

- Medications- antidepressants (if severe and not improving)

- Talking therapies – CBT

- Self help groups

- Balanced life style- regular exercise, healthy diet

SELF HELP

- Try not to blame yourself or make any significant changes in your life that can be stressful like moving house.

- Rest as much as you can, get enough nourishment, find time to have fun with your partner, let yourself and your partner be intimate if you can.

- Make friends with other couples in a similar situation.

- Find someone to talk to. If you can't try the National Childbirth Trust - they organise groups.

- If you are planning future pregnancies, go to antenatal classes and take your partner.

- Keep in touch with your GP and health visitor if you have had PND before.

WHAT CAN A PARTNER OR FAMILY DO?

- Take the time to listen

- Try not to be shocked or disappointed by the diagnosis. It can be treated

- Give practical help like shopping, feeding, changing the baby, or housework

Prognosis?

- 10-15% risk of PND if no history of depression

- After one episode, 20-40% risk of repeat episode in future pregnancy.

PUERPERAL PSYCHOSIS- EXPLANATION

What is puerperal psychosis?

Introduce the diagnosis gently (most often the partner/ actor will be upset/ angry with your diagnosis. Ask him how much is he aware of what is happening with his partner. When he tells you some of the symptoms like "behaving strangely", tell him that these symptoms are still present and then introduce the daignosis as a psychotic episode. Puerperal psychosis is a mental illness that occurs in women within 6 weeks of childbirth. It affects 1 in 500 mothers.

Causes

- **Runs in families**

- **Hormone changes**

- Risk factors:

    - **Lack of social support**

    - **Stress**

    - **Pre-existing psychiatric illness** (if there is a past history of BPAD, 50% chance of developing puerperal psychosis)

    - **Family history** of mental illness

Symptoms

- Mania

    - Full of energy and confidence

    - Stay up all night

    - Eat little

    - Irritable at times

    - Will neglect her baby because she feels that she has so many other things to do

- Depression with suicidal ideation. Rarely, a mother will kill her baby

- Schizophrenic symptoms

. Treatment

- Depending on the severity and the risk assessment, patient can be managed at home with additional support or admit to a mother and baby unit so that bonding between mother and baby will not be disrupted.

- MDT- Obstetricians, GP, midwife, health visitor

- Usually given antidepressant and antipsychotic drugs

- Cant use hormones as we don't understand the way in which the hormone changes work.

- Assess support from the family (family may need additional support)

- Voluntary groups- Association for Post Natal Illness

Recurrence

- The risk of having another puerperal psychosis is at least 1 in 5

- Half the women who suffer a puerperal psychosis never become mentally ill again

ANTIDEPRESSANTS IN PREGNANCY

- Lowest risk in pregnancy - Fluoxetine, TCAs such as amitriptyline, imipramine and nortriptyline

- SSRIs after 20 weeks- increased risk of pulmonary hypertension

- Paroxetine during 1st trimester- risk of fetal heart defects

- Venlafaxine - increased risk of toxicity in overdose, risk of high BP, increased difficulty in withdrawal

- Imipramine, nortriptyline and sertraline have lower levels in breast milk

- Citalopram and fluoxetine – high levels in breast milk

- Withdrawal or toxicity in neonates with all antidepressants

LITHIUM IN PREGNANCY

- Fetal heart defects (increased from 8 in 1000 to 60 in 1000)

- Ebstein's anomaly (increased from 1 in 20,000 to 10 in 20,000)

- Do not prescribe in 1st trimester

- If planning pregnancy and is mentally well - stop lithium

- If already pregnant on lithium and low risk- gradual reduction over 4 weeks

- If high risk- switch to antipsychotic or consider restaring Lithium in the second trimester

- If continuing to take Lithium- serum levels every 4 weeks, weekly from 36th week, and within 24hrs after childbirth

- After childbirth, adjust to lower end of serum levels

- High levels in breast milk - hence do not breast feed on Lithium

BENZODIAZEPINES IN PREGNANCY

- Cleft palate and other fetal malformations

- Floppy baby syndrome

- Do not routinely prescribe

CARBAMAZAPINE IN PREGNANCY

- Neural tube defect- risk increased from 6 in 10,000 to 20-50 in 10,000

- GI and cardiac anomalies

VALPROATE IN PREGNANCY

- Neural tube defect (spina bifida and anencephaly risk increased from 6 in 10,000 to 100- 200 in 10,000)

- Effects on child's intellect

- Polycystic ovarian syndrome in women <18yrs

- Do not prescribe in women of childbearing age

- Use adequate contraception if there is no alternative to valproate

- Stop if pregnant and consider antipsychotic

- Limit doses if it has to be used to 1g daily in divided doses with 5mg folic acid daily

LAMOTRIGINE IN PREGNANCY

- Oral cleft (risk- 9 in 1000)

- Stevens Johnson syndrome

ANTIPSYCHOTICS IN PREGNANCY

- Raised prolactin- reduced chance of conception

- Gestational diabetes and weight gain with olanzapine

- Agranulocytosis with Clozapine

- Mother on Depot- EPSE in infant

- Do not give clozapine or depot to pregnant women

CHAPTER 10   PERSONALITY DISORDERS

PREMORBID PERSONALITY

- Prevailing mood

- Hobbies

- Ability to motivate self or initiate actions

- Relationships

    - Current friends/ partners

    - Previous relationships- ability to maintain

    - Childhood- shyness at school/do they still have school friends

- Work

- Sociability

- Religious beliefs

- Coping strategies

Ask 2-3 questions for each personality trait/disorder

PERSONALITY DISORDERS

- Patterns of experience and behavior that deviate markedly from culturally accepted range in areas such as cognition, affectivity, control over impulses and gratification and handling of interpersonal relationships

- Enduring and long standing

- Pervasive and maladaptive

- Appear during childhood or adolescence

- Personal distress as a result

- Problems with occupational or social functioning

- **Paranoid**

  - How do you get on with people?

  - Do you find difficult to trust people?

  - Do you have difficulty forgiving people if they insult you?

  - Are you a very sensitive person?

- **Schizoid**

  - Would you describe yourself as a "loner"?

  - Do you have many friends?

  - How do you spend time (few activities)?

  - Emotional coldness

  - Excessive preoccupation with fantasy

- **Schizotypal**

  - Has anyone ever told you that you think oddly?

  - Do you have difficulty in thinking?

  - Do you struggle with your emotions?

- Any unusual experiences like hearing or seeing things when no one is there around?

- **Antisocial (Dissocial)**

  - What do your friends think of you? (callous unconcern for others feeling)

  - Forensic history (irresponsiblity and disregard for social rules)

  - Relationship history (incapacity to maintain relationships)

  - How do you feel if you have to queue up or stuck in traffic? How would you react if someone jumped a queue in front of you? (very low tolerance to frustration)

  - Do you feel guilty if you hurt somebody or is it always someone else's fault? (lack of guilt or remorse/marked proneness to blame others)

  - When taking antisocial history, also ask for history of persistent irritability, conduct disorder and ADHD during childhood

- **Anxious (avoidant)**

  - Are you an anxious or shy person?

  - Do you feel that you are not as good as other people?

  - Are you a worrier?

  - Do you often worry that others will criticise you?

  - Do you avoid ssocial activities because you worry that others will criticise you?

- **Dependent**

    - How much do you depend on others?

    - Do you allow others to make important decisions for you?

    - Do you feel uncomfortable or helpless when alone?

    - Do you struggle with everyday decisions?

- **Anankastic**

    - Do you keep things in an orderly way?

    - Are you a perfectionist?

    - Do you have unusually high standards at work?

    - Do you think you are pedantic about things?

- **Histrionic**

    - Are you overemotional?

    - Are you easily influenced by others?

    - Do you like being the centre of attention?

    - Are you overly concerned with your physical appearance?

- **Narcissistic**

    - Strong sense of self importance

    - Dream of unlimited success and power

    - Crave attention from other people but show few warm feelings in return

- Exploit others

- Ask for favours that you do not return

- **Emotionally unstable - impulsive type**

  - Acts without thinking (Do you ever do things impulsively and then regret them? Examples- spending habits, substance abuse, reckless driving)

  - Emotional instability (How have your relationships worked for you? How did you feel about your previous carers?)

  - Outburts of violence or threatening behaviour, particularly in response to criticism (Do you think you have a problem with temper? How often have you got in to fights?)

  - Changing goals frequently (Looking back at your life do you feel that you ambitions or goals in life have often chnaged? Has this made it difficult for you?)

- **Emotionally unstable - borderline type**

  - Impulsiveness

  - Emotional instability

  - Poor self image

  - Chnaging goals frequently

  - Internal preferences unclear

- Chronic feelings of emptiness (Do you often feel as though you are empty inside? As though you are life is meaningless? How often do you feel bored? Do you feel bored even when you are doing something you ought to enjoy?)

- Excessive efforts to avoid abandonment (How often do you feel that you are left on your own to cope with your problems? Do you ever feel that people would leave you? What happens when you feel this way?)

- Series of suicidal threats or acts of deliberate self harm

PERSONALITY DISORDER - EXPLANATION

What is a personality disorder?

- Personality is a collection of ways that we think, feel and behave that makes each of us an individual. People who have personality disorder have parts of their personality that make it hard for you to live with themself and others.

- 1 in 10 have a personality disorder.

- It can lead to problems in making or keeping relationships, problems with people at work, problems with friends and family, getting into trouble with the law and finding it difficult to control your feelings or behaviour.

- 40- 70% of people on a psychiatric ward , 30-40% of psychiatric patients being treated under CMHT,10- 30% of patients who see their GP

Causes

- **Upbringing**: More at risk of developing a PD if there is history of physical or sexual abuse in childhood, violence in the family, family substance misuse.

- **Early behavioural problems** eg conduct problems

- **Brain problems**- more demonstrable with dissocial personality

- Worsened by substance misuse, relationship and money problems, associated problems such as anxiety, depression and other mental health problems

Treatment

- Psychological
  - Counselling
  - Dynamic psychotherapy- looks at how past experiences affect present behaviour
  - Cognitive therapy- changes unhelpful patterns of thinking
  - Cognitive analytical therapy- recognisess and changes unhelpful patterns in relationships and behaviour
  - Dialectical behaviour therapy- combination of cognitive and behavioural therapies and techniques from Zen Buddhism. It involves individual therapy and group therapy
  - Treatment in a therapeutic community- this is a place where people with long-standing emotional problems can go to (or sometimes stay) for

several weeks or months. Most of the work is done in groups. You learn from getting on or not getting on with other residents. The staff and other residents help you to get through such problems and learn from them.

- Medications
  - Antipsychotic drugs (usually at a low dose) reduces the suspiciousness of the three cluster A personality disorders (paranoid, schizoid and schizotypal). It can help with borderline personality disorder if people feel paranoid, or are hearing noises or voices.
  - Antidepressants- can help with the mood and emotional difficulties that people with cluster B personality disorders have. Some of the SSRIs can help people to be less impulsive and aggressive in borderline and antisocial personality disorders. It can reduce anxiety in cluster C personality disorders (obsessive-compulsive, avoidant and dependent).
  - Mood stabilisers- Lithium, carbamazepine, and sodium valproate can also reduce impulsiveness and aggression.

- Admission to hospital usually happens only as a last resort and for a short time.
- Aggression and impulsiveness seem to reduce in 30s and 40s. Schizotypal personality disorder can develop into schizophrenia.

PATHOLOGICAL GAMBLING

- Frequent, repeated episodes of gambling which dominates the individual's life to the detriment of social, occupational and family values and commitments

- May put their jobs at risk, acquire large debts, and lie or break the law to obtain more money or evade payment of debts

- Personal distress due to gambling

- Affects personal functioning

- Intense urge to gamble and cant stop by own will

- Preoccupied with thoughts and images of gambling which often increases at times when life is stressful

- Rule out normal gambling, gambling by manic patients, gambling by sociopathic personalities

- Treatment - SSRIs, naltrexone, CBT

PATHOLOGICAL FIRESETTING (PYROMANIA)

- Repeated fire-setting without any obvious motive such as monetary gain, revenge, or political extremism

- An intense interest in watching fires burn

- Reported feelings of increasing tension before the act and intense excitement immediately after it has been carried out

- Preoccupied with thoughts of firesetting or of circumstances around the act

- Rule out deliberate firesetting, conduct disorder, firesetting in dissocial personality disorder, firesetting in schizophrenia, organic disorders

- Treatment- CBT

PATHOLOGICAL STEALING (KLEPTOMANIA)

- Repeated failure to resist impulses to steal objects that are not acquired for personal use or monetary gain

- The objects may instead be discarded, given away, or hoarded

- Increasing tension before and relief afterward

- The theft is a solitary act, not carried out with an accomplice

- May express anxiety, despondency and guilt between episodes of stealing from shops

- Rule out normal shoplifting, organic mental disorder, depressive disorder with stealing

- Treatment - SSRIs, Psychotherapy (CBT, family therapy)

TRANSSEXUALISM

- Wants to live and be part of opposite sex

- Transsexual identity for at least 2 years

- Not due to other mental illness

DUAL ROLE TRANSVESTITISM

- Wears clothes of opposite sex for temporary membership of group

- No sexual motivation for cross dressing

- No desire to change sex

GENDER IDENTITY DISORDER OF CHILDHOOD

- Persistent and intense distress about being boy/girl

- Wears masculine clothing or asserts that she will grow penis, rejects urinating in sitting position, asserts that she doesn't want to grow breasts or menstruate

- In boys - believe that they will grow up to be a woman, penis will disappear, it would be better not to have a penis or testes.

- Has not reached puberty

- Present for at least 6 months

EXHIBITIONISM

- Recurrent tendency ro expose genitalia to unsuspecting strangers of opposite sex

- No intention to have sexual intercourse with these people

PAEDOPHILIA

- Persistent or predominant preference for sexual activity with a prepubescent child

- Recurrent sexual urges and fantasies which are acted upon

- Individual is at least 16 and 5 years older than child

CHAPTER 11   FORENSIC PSYCHIATRY

INDECENT EXPOSURE

- Demographics:

    - Employment

    - Marital status

    - Current home situation

- Incident:

    - Patients account

    - Circumstances- confusion/ intoxication

    - Motivation

    - Understanding of implications- is it keeping in character?

    - Insight

    - Remorse

    - Previous offences of a similar type

    - Previous convictions for a similar offence

    - Personal history of abuse

- Differential diagnosis:

    - Dementia- FTD, AD

    - Confusional state- organic, medication, substances

    - Psychotic

    - Affective

- Sexual disorder- exhibtionism

- Substance misuse/ dependence

- Risk assessment

  - Nature of offence

  - Previous offences and convictions

  - Context of offending- disturbed mental state or not

  - Insight, remorse, deterrent effect of arrest

FITNESS TO PLEAD

- Index episode- explore in detail

- Understand the charge- what and why

- Difference between pleading guilty and not guilty

- Instruct counsel- Do you know what a solicitor is and what he does? Would you be able to represent yourself in court?

- Able to follow court proceedings

- Challenge the juror- (knowing somebody in the jury could act against them) If you knew that someone in the jury would act against you, could you tell the judge?

- Previous history of court attendance

- Screen for symptoms of other illness- psychosis, mania/hypomania, depression, learning disability, drugs and alcohol

- Current medications

- Forensic history

- Past psychiatric history

- Social circumstances

- Premorbid personality

PAEDOPHILIA- HISTORY

- Details of alleged incident

- Friends or acquaintances accused of similar acts (paedophile ring)

- Fantasies- involving adults, children, violence

- Internet surfing activities- chat rooms, child pornography

- Access to or possession of child pornographic materials

- Access to children or young poeople

- Attitiude of patient to sex with children

- Relationship history

- Choice of occupation to facilitate access

- Previous similar allegations

- Forensic history

- Factors important in management

    - Cognitive disortion- rationalieses victim is consenting

    - Denial

- Co-morbidities- alcohol, LD, PD, Psychosis, hypomania, temporal lobe pathology

- Motivation for treatment – non compliance – high risk of repeating

- Management

  - Police need to establish if there is enough evidence to charge

  - Social services involvement with child and family

  - Full psychiatric assessment to exclude any other mental health problems

  - May need psychological assessment if convicted to assess attitudes towards his behaviour to children and victim empathy

  - CBT and SOTP (forces paedophile to confront and rectify distorted thinking)

  - Pharmaceutical treatments- estrogens, neuroleptics, antiandrogens, medoxyprogesterone acetate and SSRI - affects the testosterone levels and decreasing sexual drive and offending.

ASSAULT ON STAFF

- Antecedents

  - Events preceeding one week

  - Affective/psychotic symptoms

  - Substance misuse

  - Triggering factors

  - Any changes

- Behaviour

  - Exact details of assault

  - Extent of injury

  - Slap/punch/kick – how many times

  - Use of weapons

- Consequences

  - Did he stop assaulting/was he dragged off

  - Was he given any medications

- Current mental state- assess for any mental illness including any remorse

- Past history of aggression

- Relevant social and personal history

- Ask the nurses view of incident

- Summarise patient story

- Management

  - Make staff aware of the risk

  - Predictive risk factors- low frustration tolerance, impulsivity, tense and angry facial expressions, restlessness, increased volume of speech

  - Strategies to minimize frustration including 1 to 1 time with nurse, more personal space, alternative activities to engage during inpatient stay

RISK OF OFFENDING

- Index episode

    - Full detail of the incident

    - Drugs/ alcohol

    - Provocation

    - Triggering factors

- Previous violence- severity of last offence

- Psychopathy

    - Impulsivity

    - Callousness

    - Lack of empathy

    - Criminal versatility

- History of substance misuse

- Any current mental illness/ personality disorder

- Past psychiatric history- assess response to treatment

- Past medical history

- Personal history

    - Traumatic childhood, violent father, dometic violence

    - Bedwetting, firesetting, cruelty to animals

- Social history

- Forensic history

    - Young age at 1st violent incident

- Prior supervision failure

- Current mental state

  - Guilt, denial, paranoid symptoms, lack of insight, ongoing thoughts of violence, negative attitude to law and other authorities

- Personality- impulsivity, self control, coping with stress

HCR 20

Historical	Clinical	Risk management
H1 previous violence	C1 lack of insight	R1 plans lack feasibility
H2 young age 1st violent incident	C2 negative attititudes	R2 exposure to destabilization
H3 relationship instability	C3 active symptoms of mental illness	R3 lack of personal support
H4 job problems	C4 impulsivity	R4 non-compliance with remediation attempts (treatment)
H5 substance misuse	C5 unresponsive to treatment	R5 stress (reaction to stress)
H6 major mental illness		
H7 psychopathy (impulsivity, criminal versatility, callousness, lack of empathy)		
H8 early maladjustment		
H9 personality disorder		
H10 prior supervision failure – breaking bail conditions		

CHAPTER 12   LEARNING DISABILITY

MENTAL RETARDATION (ICD 10 terminology)

A condition of arrested or incomplete development of the mind, which is especially characterized by impairment of skills manifested during the developmental period, skills which contribute to the overall level of intelligence, i.e. cognitive, language, motor and social abilities. Retardation can occur with or without any other mental or physical condition.

MILD MENTAL RETARDATION

- IQ 50- 69 (mental age 9-12years)
- Likely to result in some learning difficulties at school
- Many adults will be able to work and maintain good social relationships

MODERATE MENTAL RETARDATION

- IQ 35-49 (mental age 6-9 years)
- Likely marked developmental delays in childhood and most develop some degree of independence in self care and acquire adequate communication and academic skills
- Adults will need varying degrees of support to live and work in the community

SEVERE MENTAL RETARDATION

- IQ 20-34 (mental age 3- 6 years)

- Likely to need continuous support

PROFOUND MENTAL RETARDATION

- IQ under 20 (mental age below 3 years)

- Severe limitation in self care, continence, communication and mobility

PREGNANCY & LD STATION

Social services is there to perform assessments of the mother, living arrangements and home situation as possible sources of support. Baby being taken into care is an extreme situation. It is more likely that they will discuss other options like additional support in the home setting. Mother's concerns can be discussed further with the community nurse with patient's permission.

CHAPTER 13   CHILDHOOD DISORDERS

DEVELOPMENTAL HISTORY

- Pregnancy

- Labour

- Breast feeding

- How was the baby

- Weaning

- Ability to grip things (palmar grip 6 months)

- Sit up without help - 9 months

- Crawling- 10 months

- Single words - 11 months

- 2-3 words with meaning - 12 months

- Walking - 12 months

- Toilet trained - 18 months

- Dry by day - 2 years

- Dry by night - 3 years

- Can ride tricycle - 3 years

CHILDREN WITH PSYCHOSIS

- Olanzapine and Risperidone effective

- At increased risk of EPSEs

- Early intervention Team

- Social skills training/CBT

CHILDHOOD AUTISM

Development disorders that are due to the abnormality in the way the brain develops. Autism shows itself in the first 3 years of life with difficulties in socializing, communication and behaviour. It is 3-4 times more common in boys.

ASPERGER'S SYNDROME

- Asperger's syndrome is a less well defined condition and may be not be recognized until after a child starts school

- They have the same kind of qualitative abnormality of reciprocal social interaction that typify autism, together with restricted stereotyped repetitive interests and activities

- No general delay or retardation in language or cognitive development

- More common in boys

- Strong tendency for the abnormality to persist into adult life

- Psychotic episodes occasionally occur in early adult life

Socialising difficulties

- Tend to ignore other people or appear insensitive to others needs, thoughts or feelings.

- Do not make usual eye contact or use facial expression in social situations.

- Less likely to use gestures such as pointing to communicate

- Find it difficult to co operate, share or take turns with others

- Prefer to play alone and show no interest in imaginative play

- Get on best with understanding adults

- Socializing with other children and forming friendships is hard for them

Communication difficulties

- Not being able to communicate properly (usually the first cause of concern)

- Language problems- both in understanding and speaking

- More severely affected- never learn to speak, or communicate in other ways. If they begin to talk, they may simply echo what they just heard or speak in an odd way

- Relative lack of creativity and fantasy in thought process

- Lack of emotional response to other people's verbal and nonverbal overtones

- Difficulty in modulation of voice tone

Unusual behaviour

- Restricted, repetitive, stereotyped patterns of behaviour, interests and activities

- Prefer familiar routine and tend to resist change which they find difficult and unpleasant

- Unusual interests such as maps or electronic gadgets

- Very sensitive to tastes smells and sounds

- Odd body movements such as hand flapping or finger twiddling

- Any attempt to stop these activities and interests can cause furious protest and distress- tantrum, hyperactive or injure themselves

- Some have special talents for drawing, music or calculation

- Specific attachment to unusual, typically non soft objects

- Insist on performance of particular routines in rituals of a nonfunctional character

Aetiology

Little is known

- Genetic (increased in Down's and fragile X)

- Obstetric complications

- Toxic agents like lead

- Pre and postnatal infections (increased in rubella)
- Associated with neurological disorders such as tuberous sclerosis

Assessment

- Multi-disciplinary approach- Speech and Language Therapist (SALT), educational psychologist, CAMHS, school observation
- Full and detailed neurodevelopmental history, family history and psychological testing
- Rating scales- Autism Diagnostic Observational Schedule (ADOS)

Treatment

- Careful consideration needs to be given to family factors relevant to the planning of traetment and support that will be offered to the family
- Parents should be reassured that their behaviour has not caused the condition and that they can help their child
- No medications and no cure
- Treatment approaches are most successful for non speech behaviour problems
- Behavioural and educational psychology, SALT- require accurate appraisal of the child's skills and difficulties, conducted in natural environment
- Parents have a very important role acting as "co-therapists"

- Generally a series of small steps change the child's behaviour with the minimum of distress
- Education with speech and language offers the most effective way of making sure that the child with autism reaches their full potential
- Dpending on what resources are needed, a child may do best in a mainstream school that provides support for children with communication difficulties, or in a special school for children with autism
- Medications
    - Antipsychotics for stereotypies
    - SSRIs for compulsive and self harming behaviour and depression and anxiety

- Children with Asperger's often also require individual help, usually in a mianstream setting
    - Need help with social rules, how to manage conflict and upset feelings
    - Need feedback on how they are getting on with other people
    - Unstructured situations, such as break and lunch times, can be particularly difficult times for these children
    - Can be vulnerable to bullying especially in mainstream schools
- Social services may be able to offer practical support
- Autistic society

RETT'S SYNDROME

- Onset between 7 and 24 months
- Characteristic loss of purposive hand movements and acquired fine motor manipulative skills
- Loss/partial loss or lack of development of language
- Sterotyped tortous wringing of hands with arms flexed in front of chest or chin
- Stereotypic wetting of the hands with saliva
- Lack of proper chewing of food
- Episodes of hyperventilation
- Almost always a failure to gain bowel and bladder control
- Excessive drooling and protrusion of the tongue
- Loss of social engagement
- Typically retain a social smile, looking at or through people, but not interacting socially
- Stance and gait broad based, hypotonic muscles, poorly coordinated trunk muscles, scoliosis, kyphosis
- Epileptic fits before age 8
- DSH and complex stereotyped preoccupations with routines are rare
- Progressive motor deterioration confirms the diagnosis

CHILDHOOD DISINTEGRATIVE DISORDER

- Normal development for 2-3 years followed by loss of acquired motor, language and social skills between ages of 3 and 4 years

- Stereotypies and compulsions are common

- Cause is unknown

- Poor prognosis

ADHD – HISTORY

- ADHD is a developmental disorder characteristed by over activity, impulsivity and inattention in more than one situation- home, school and socially

- Affects 1-2% of children

- Onset before age 6 years

- Priority over conduct disorder

- More common in males (3:1)

- More than one situation – home, school, clinic

Impaired attention

- Prematurely breaking off from tasks

- Leaving activities unfinished (lose interest and get diverted easily)

- Fails to give close attention

- Fails to sustain attention

- Often appears not to listen

- Often fails to follow through on instructions/ finish schoolwork/ chores/ duties

- Impaired organizing tasks/activities

- Avoids tasks requiring sustained mental attention

- Loses things necessary for certain tasks

- Easily distracted

- Forgetful

Overactivity

- Excessive restlessness especially in situations requiring calm (running, jumping, getting up from seat)

- Excessive talkativeness, noiseness, fidgeting and wriggling

- Unduly noisy in playing

- Persistent excess motor activity

Impulsivity

- Blurts out answers

- Often fails to wait in lines

- Interrupts others

- Talks excessively without appropriate response to social constraints

Other features

- Disinhibited in social relationships

- Unpopular with other children, become isolated

- Associated reading difficulties

- Secondary complications- dissocial behaviour, low self esteem

Aetiology

- Genes- increased level of conduct disorder and substance misuse in parents

- Chemical imbalance of noradrenaline and dopamine

- Stress

- Family dysfunction

- Poor attachement

Co morbidity

- Specific learning disorders- 60%

- Conduct and oppositional defiant disorder- 40%

Exclude differentials

- Home - marital conflict, violence

- School - bullying

- Behaviour- conduct/ oppositional defiant disorder

- Development- ASD, speech and language problems, tourettes

- Psychiatric- depression, anxiety, psychosis, substance misuse

- Medical - hearing, fragile X, thyroid, epilepsy, history of rheumatic fever

Management

- Diagnosis is made only by specialists- Paediatrician or Psychiatrist

- Child and family interviews- developmental history

- School and home observations, Conners questionnaires

- May need assessment by OT, SALT, Educational Psychologists, Special Education Need Coordinators (SENCO)

- Balanced diet, regular exercise, good nutrition

- Eliminating artificial colours and additives is not recommended or dietary fatty acid supplements. Parents can keep diary to see if any particular foods that affect behaviour

- In primary care - watchful waiting for up to 10 weeks and referral to parent training but if there is moderate impairment- refer to CAMHS

- Pre-school
    - Parent training- 8-12 sessions including role play and homework
    - Drugs not recommended

- School age

- Moderate impairment should have parents referred to group parent training and/ or group treatment program (CBT or social skills training)
- Individual psychological intervention (CBT/ social skills)
- Persisting problems or when non drug intreventions have been refused- offer medications

- School age with severe ADHD
  - Medications and group based parent training
- Adult with ADHD
  - Drug treatment first line choices as for child
  - Consider group or individual CBT if problems with meds or response to meds
  - Continue treatment for as long as beneficial
- Drug treatment
  - Methylphenidate for no cormorbidity or with additional conduct disorder
  - Methyphenidate or atomoxetine in the presence of tics, Tourette's, anxiety, stimulant misuse
  - Atomoxetine if Methylphenidate ineffective or intolerant
  - Do not use antipsychotics
  - Full history and physical examination

- History of exercise syncope, shortness of breath, blood pressure, height and weight
- Family history of cardiac disease
- ECG if any cardiac concerns
- Risk assessment for substance misuse

METHYLPHENIDATE

- Stimulant
- More effective at treating hyperacitivity than inattention
- Onset of action within 30 mins, lasts 3 hours, sustained release lasts 6-10 hours
- 0.2mg/kg/dose max 0.7mg/kg
- Titrate slowly over 4-6 weeks, until no further improvement
- Not addictive, don't make them high and don't cause sedation
- Side effects
  - Headache, stomach ache, dysphoria, nervousness- wears off spontaneously or responds to dose decrease
  - Appetite suppression/ insomnia
  - Raised BP, dizzy, rash, nausea – BP every 3/12
  - Repetitive activities/stereotypies- disappears when dose reduced

- Height- reduces height by 2 cm for every 3 years taking medication; check height every 6 months; drug holiday to allow growth to catch up in school holidays
- Weight - check every 3- 6 months
- Seizures exacerbated

- Consider modified release (once daily) preparation to improve adherence, decreased stigma (need not be taken in school), decreased problems with storing and administering controlled drug in school

ATOMOXETINE

- Nor Adrenaline Reuptake Inhibitor (NARI)
- 0.5 mg/kg/day
- Affects height and weight
- Side effects
    - Agitation, irritability, suicidal thoughts, behaviour changes – especially after dose change or in initial months
    - Dysmenorrhoea, erectile dysfunction, ejaculatory dysfunction
    - Seizures exacerbated

PARENT TRAINING

- Improves child management skills and therby reduce family stress and children's negative behaviour, thereby improves prognosis by decreasing the likelihood of future substance abuse and antisocial behaviour
- Delivered by supervised, appropriately trained therapists
- Involves behavioural programme and relationship programme
- Behavioural programme- teaching parenting skills to reduce and cope with behaviour
- Relationship programme- helps parents to understand both their own and their child's emotions and behaviour and improve their communication with the child
- Structured
- Ideally 8-12 sessions
- Enable parents to identify their own parenting objectives
- Incorporate role play during sessions and homework

Good parenting tips

- Consistent
- Calm
- Give lots of praise for good behaviour
- Rewards for good behaviour

- Brief periods of time away from other people if negative behaviour becomes too much
- Planning ahead
- Involve child
- Be clear with child
- Be realistic

Outcome

- 20% develop dissocial personality traits
- 15-20% develop substance misuse
- High rates of suicidaility, poor self esteem and unemployment
- Inattention often persists
- In adulthood- 20-30% full ADHD syndrome, 60% have one or more symptoms

CONDUCT DISORDER

Occurs for at least 6 months

Causes

- Difficult temperament
- Learning or reading difficulties
- Depressed
- Bullied/abused

- Hyperactive

- Rule out dissocial personality disorder, schizophrenia, mania, depression, pervasive development disorder, hyperkinetic disorder

- Frequent/severe tantrums

- Argues with adults

- Refuses adults requests

- Deliberately does things that annoy others

- Blames others for own mistakes

- Easily annoyed

- Angry/resentful

- Spiteful/vindictive

- Often lies or breaks promises to obtain goods or favours

- Frequently intiates physical fights

- Has used weapon that can cause serious physical harm

- Often stays out after dark (starting before age 13)

- Physical cruelty to other people

- Physical cruelty to animals

- Deliberately destroys the property of others

- Deliberately sets fire with risk/intention of causing serious damage

- Steals objects of non trivial value

- Frequently truant from school beginning before age 13

- Has run away from home

- Commits crime involving comfrontation of victim

- Forces another person into sexual activity

- Frequently bullies others

- Breaks into someone elses car/house

OPPOSITIONAL DEFIANT DISORDER

- Conduct disorder occurring in younger children

- Same criteria as above

- Usually younger than 9/10 years old

- Absence of behavior that violates rights of others

- Definance has a provocative quality- rude, hostile, resistance to authority

SEPARATION ANXIETY DISORDER OF CHILDHOOD

- Occurs before age 6

- Does not meet the criteria for GAD

- Lasts at least 4 weeks

- Unrealistic persistent worry about possible harm befalling major attachment figures

- Unrealistic persistent worry that untoward event will separate child from major attachment figure

- Reluctance to go to school because of above fears

- Reluctance to go to sleep because of above fears

- Persistent inappropriate fear of being alone or without attachement figure

- Repeated nightmares about separation

- Repeated occurrence of physical symptoms

- Excessive recurrent distress in anticipation of separation

BULLYING

- Bullying is when a child is picked on by another child or group of children

- Includes teasing, calling someone names, threatening or harrassing them

- Sometimes it getys physical- taking the child's belongings or even pushing or attacking them

- 1 in 4 primary school pupils and 1 in 10 secondary school are bullied

- Can make them lack confidence, find it hard to face going to school, may feel depressed and even suicidal

- Causes- children who are more aggressive already may pick on children who are different in some way like quiet/ shy/ alone at playtime

- Be open to the possibility that your child might be bullied

- Listen to your child

- Take them seriously

- Do not blame the child

- Do no promise to keep bullying a secret

- Make them realise its not their fault

- Talk with your child and work out ways of solving the problem

- School has a responsibility to sort it out – should have an anti bullying policy

SCHOOL REFUSAL

- Child may be too anxious to go to school

- Worrying about going to school may make them vaguely unwell with sickness, headaches, tummy aches although usually no cause can be found

- Worse on weekday mornings

- Disappear later in the day

- May be fearful about leaving the safety of their home

- Clingy and lacking in confidence

- Once they get involved in lessons they may find that they enjoy school.

- Causes

  - Family problems

  - Changes in the family (birth of a brother/ sister)

  - Illness in the family

  - Bullying at school

- Talk to them and support them

- Work with child's teachers to encourage your child to go back to school as quickly as possible

- Sort out any underlying problems like bullying

- May need to involve education welfare officer or educational psychologist

CHAPTER 14   PHYSICAL EXAMINATIONS

- Briefly ask about the presenting complaints

- Tell them that you will be as gentle as possible during the examination

- Always mention chaperone

- Use alcohol gel

- If you dont complete the station in time, mention the remaining examination that you would have liked to do

- At the end of the task, tell the patient the need for further investigations and possible referral to your physician colleagues

- Do not forget to thank the patient

CARDIOVASCULAR SYSTEM EXAMINATION

- Hands – cyanosis, clubbing, splinter haemorrhages, palmar erythema

- Pulse – rate, rhythm, volume, character

- BP

- Eyes- pallor and jaundice

- Tongue- cyanosis

- JVP at 45 degrees

- Inspection

    - Shape of the chest

- Scars

- Deformity

- Redness

- Apex beat

- Palpation

  - Apex beat

  - Thrills

  - Parasternal heave

- Auscultation

  - Tricuspid

  - Pulmonary

  - Aortic

  - Mitral

  - Bases of lungs

- Hepatomegaly

- Peripheral oedema

RESPIRATORY SYSTEM EXAMINATION

- Hands- pallor, cyanosis, clubbing

- Pulse

- BP

- Eyes- pallor, jaundice

- Tongue- pallor, cyanosis

- Cervical lymph nodes

- Inspection

    - Shape of the chest

    - Respiratory rate

    - Movement of chest wall

    - Use of accessory muscles

    - Scars

    - Redness

- Palpation

    - Trachea

    - Movements- anterior, posterior and lateral chest expansion

    - Vocal fremitus

    - Apex beat

- Percussion

- Auscultation

ABDOMINAL EXAMINATION

- Hands- cyanosis, pallor, clubbing, palmar erythema, dupytrens contracture

- Pulse

- Eyes- jaundice and pallor

- Mouth- cyanosis and foetor hepaticus

- Cervical lymphadenopathy

- Chest- gynaecomastia/spider naevi

- Inspection

    - Size

    - Shape

    - Swelling

    - Scars

    - Pulsations

    - Caput medusa

- Palpation

    - General tenderness

    - Liver, kidneys, spleen

    - Fluid thrill

- Percussion

    - Liver dullness

    - Fluid shift

- Auscultation

    - Bowel sounds

    - Aortic and renal bruits

- Mention PR examination

EXTRAPYRAMIDAL SIDE EFFECTS EXAMINATION

- Ask for any dentures or sweets inside the mouth

- Ask briefly about abnormal movements

- Move head up and down (acute dystonia)

- Observe for abnormal movements in sitting position for 15 seconds- hands on knees, feet flat on the floor, legs slightly apart

- Open mouth twice and protrude tongue twice

- Ask the patient to tap thumb with each finger and observe to see if movements are slow

- Flex and extend arms (cogwheeling)

- Standing- extends arms out with palms down

- Check for gait

- Check Romberg's sign

CRANIAL NERVE EXAMINATION

- 1st nerve (olfactory)- Have you noticed any difficulties in the way you smell things?

- 2nd nerve (optic)
    - Visual acuity- reading a near vision chart at 30 cms
    - Visual fields- focus on bridge of your nose while testing eyes
    - Light reflex- direct and indirect

- Accommodation- look at a point far away and then at finger that is 10cm away

- Mention colour vision (Ishihara plates are not provided)

- 3rd, 4th and 6th nerve (Occulomotor, Trochlear, Abducent)

  - Eye movements- H pattern (ask patient to report any double vision, note nystagmus/saccadic eye movements)

  - Convergence (look into distance and then focus on finger brought in from 50cm to limit of convergence)

- 5th nerve (Trigeminal)

  - Sensory- test all three divisions in the face using a wisp of cotton

  - Motor- feel masseters and temporalis when teeth clenched

- 7th nerve (Facial)

  - Motor- raise eyebrows, blow out your cheeks

  - Sensory- any problems with taste

- 8th nerve (Vestibulocochlear)

  - Make noise by rolling fingers near both the ears

  - Rhinnes test

  - Webers test

- 9th and 10th nerve (Glossopharyngeal and vagus)

  - Open mouth and say "ahh"

  - Cough

- 11th nerve (Accessory)

- Shrug shoulders against resistance

- Turn head left or right against resistance

- 12th nerve (Hypoglossal)- Stick tongue out and move it side to side

THYROID EXAMINATION

- Hands- sweating, warmth, tremors

- Nails- clubbing/cyanosis/pallor

- Pulse

- BP

- Inspection

    - Eyes- exompthalmous and eye movements, lid retraction or lid lag

    - Swelling of the gland

    - Protrude tongue

    - Scars, sinuses, visible pulsation

- Palpation

    - Of the gland from front and behind

    - Carotid pulses

    - Cervical lymphadenopathy

- Percussion- clavicle

- Auscultation for bruit

- Reflexes

- Ankle oedema

Hyperthyroidism

- Weight loss
- Heat intolerance
- Sweating
- Diarrhea
- Tremors
- Irritability
- Anxiety
- Emotional lability

Hypothyroidism

- Tiredness
- Lethargy
- Weight gain
- Constipation
- Intolerance to cold
- Menorrhagia
- Hoarse voice
- Depression
- Dementia

- Myalgia
- Coarse thinning hair
- Yellowish skin tint
- Slow relaxing reflexes

UPPER LIMB NEUROLOGICAL EXAMINATION

- Inspection
    - Wasting
    - Fasciculation
- Tone
    - Gently move wrist, elbow, shoulder
- Power
    - Shoulder shrug
    - Shoulder abduction
    - Shoulder adduction
    - Elbow flexion
    - Elbow extension
    - Wrist extension
    - Finger grip
    - Thumb adduction
    - Thumb abduction
- Co-ordination

- Dysdiachokinesis

- Finger nose

- Reflexes

    - Biceps

    - Triceps

    - Suppinator

- Sensation

    - Light touch

    - Vibration

    - Proprioception

    - Pain and temperature (mention only)

LOWER LIMB NEUROLOGICAL EXAMINATION

- Inspection

    - Wasting

    - Fasciculation

- Tone

- Test for clonus

- Power

    - Hip flexion

    - Hip extension

    - Knee extension

- Knee flexion

- Dorsiflexion

- Plantar flexion

- Co-ordination- Heel shin test

- Reflexes

  - Knee

  - Ankle

  - Plantar

- Sensation

  - Light touch

  - Vibration

  - Proprioception

  - Pain and temperature (mention only)

ALCOHOL MISUSE PHYSICAL EXAMINATION

- Hands- Jaundice, palmar erythema, clubbing, dupytrens contracture, flapping or withdrawal tremors

- BP & Pulse

- Eyes- Icterus, pallor, eye movements

- Face- Parotid enlargement, facial flushing

- Inspection

  - Shape of the abdomen

- Spider naevi

- Scars, sinuses, masses, pulsations, hernia

- Palpation

  - Tenderness

  - Hepatomegaly

  - Splenomegaly

  - Kidney enlargement

- Percussion

  - Fluid thrill

  - Shifting dullness

- Auscultation

  - Bruit

  - Bowel sounds

- Neurological examination

  - Power and cogwheeling

  - Light touch sensation arms and legs

  - Gait

- Peripheral oedema

EATING DISORDERS EXAMINATION

- Skin

  - Signs of dehydration

- Purpura

- Lanugo

- Hands

  - Pallor

  - Brittle nails and hair

  - Callouses- Russell's sign

- Pulse & BP

- Eyes - anaemia

- Mouth

  - Conjunctival haemorrhages

  - Dehydration

  - Damaged enamel of teeth

  - Salivary glands

- Cardiovascular system

  - Precordial examination

  - Heart sounds (mitral valve prolapse)

- Gastrointestinal system

  - Abdominal distension

  - Tenderness

- Peripheral neuropathy

- Peripheral oedema

- Check height and weight

- Squat test- rise from squatting position unaided

OPIATE WITHDRAWAL EXAMINATION

- History

  - Cravings, withdrawal symptoms

  - Joint pain

  - Muscle pain

  - Abdominal cramps

  - Diarrhea or vomiting

  - Fever

- Skin examination

  - Sweating

  - Gooseflesh

  - Scratch marks, abrasions, bruises

  - Dehydration

  - Injection sites

- Eyes

  - Dilated pupils

  - Watery eyes

  - Jaundice

- Face

  - Rhinorrhoea

- Yawning
- Blood sugars
- Cardiovascular system
    - Hypo/hypertension
    - Tachycardia
    - Murmurs
- Respiratory system
    - Tachypnoea
    - Signs of LRTI
- Abdomen
    - Hepatosplenomegaly

Neurological
    - Wasting
    - Tone
    - Power
    - Reflexes
    - Coordination
    - Gait
    - Sensory system

CHAPTER 15   PSYCHOTHERAPY

- Talking therapy

- Indications- Severe emotional symptoms which can be understood in psychological terms

- Contraindications- Drug/ alcohol dependency, suicidal, psychosis, severe depression

- Psychotherapy is effective in reducing symptoms, improving relationships and self-esteem Benefits may not be immediate and may feel worse in the beginning

- Medication continues and can access GP/CPN in event of an emergency

- Psychotherapy records are kept separate

- Summary of the therapy is given to GP

- All therapists undergo constant supervision

- Psychotherapists are not always doctors - can be nurses, psychologists with specialist training in psychotherapy

- Therapists offer guidance - ultimate responsibility of change is with the patient

- Known to help people with a variety of problems but not everyone gains this benefit.

COGNITIVE BEHAVIOURAL THERAPY

- CBT is one of the talking therapies

- It looks at how we think about yourself, the world and other people and how what we do affects our feelings and actions

- CBT can help you change what you think (cognitive) and what you do (behaviour)

- It is based on the principle that how we feel is not because of the events but how we percieve the events

- It has been found to be helpful in Anxiety, Depression, Panic, Agoraphobia and other phobias, Social phobia, Bulimia Nervosa, Obsessive compulsive disorder, Post traumatic stress disorder and Schizophrenia

- Explain how situation interacts with thoughts, feelings, sensations and behaviour with a diagram citing an example

- Structured session

- Individual or group

- Each session lasts 50 mins once a week

- Usual course lasts between 5- 20 sessions

- With the therapist you breakdown each problem (to aid this he might ask you to keep a diary of thoughts,feelings and behaviours), look at the thoughts, feelings and actions and emotions and see if they are helpful or not and then work out what might be more helpful. You then put these into practice in daily life. So you might question a self critical thought and replace it with a positive one or recognise youre about to do something unhelpful and do something helpful instead

- Homework

- The therapist will also ask you questions about your past life and background. Although CBT concentrates on the here and now, at times you may need to talk about the past to understand how it is affecting you now

- Continue medications

- Can access GP and CPN in emergency

- Therapist undergoes regular supervision

- Can prevent relapse because eventually you will be able to use what you have learnt to sort out your problems

- It is as good as antidepressants for moderate depression

- Most effective psychological treatment for moderate to severe depression

- Can be hard to get the hang of it at first if you cant concentrate or feel low. This in turn can make you feel disappointed or overwhelmed. It can also sometimes be difficult to talk about the symptoms of depression and anxiety

COGNITIVE DISTORTIONS

Early experiences

Core beliefs (or schemas) – to succeed in life you must work harder than others

Dysfunctional assumptions/rules – if I don't get all A's then I'm a failure

Critical incident – trigger core beliefs

Sets off negative automatic thoughts

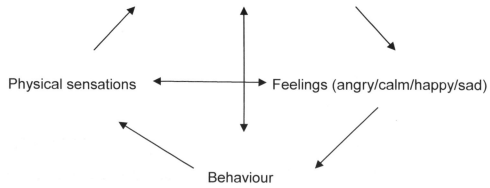

Physical sensations   Feelings (angry/calm/happy/sad)

Behaviour

Cognitive distortions

- Minimization:

  - Performance is underestimated

  - Do you often ignore or overlook the positive or good things that happen?

  - E.g.: Girl makes one error in an essay and thinks that the paper is ruined

- Magnification:

  - Errors are overestimated

  - Do you often blow up or magnify things that go wrong?

  - E.g.: Customer service agent only notices complaints not positive comments

- Overgeneralization

  - Reaching a general conclusion on the basis of a single event and making wide generalisations

  - Do you often feel that single negative events seem to snowball in to something bigger?

  - E.g.: Man does not get job and believes he will never get a job and nobody will take him

- Selective abstraction (mental filtering/ discounting the positive) – focusing on a specific detail and ignoring the more important features of a situation

- Personalisation

- Arbitrary inference

- All or none/ Dichotomous thinking

- Emotional reasoning

- Labelling

- Catastrophising

Man with considerable success at work – anyone can do it

Personalization - relating external events to oneself in an unwarranted fashion – blame self for things that you have no control over

Arbitrary inference (Jumping to conclusions) - reaching a conclusion when there is no evidence for it and even some against it (mind reader – how do you know what others are thinking/ fortune teller – how can you predict things)

Man with indigestion after spicy meal goes to see doctor and says i must have cancer. My friend didnt answer the doorbell he must be avoiding me.

Dichotomous thinking /All or nothing thinking- "Black and White thinking"; the tendency to see events as completely good or completely bad;

Woman has ruins special dinner – thinks i have ruined everything

238

Emotional reasoning – our emotions become very strong and cloud the way in which we think and see things. What we think depends on how we feel, not on what actually happens

I feel useless so i cannot be good at anything

Labelling (yourself) – you attach a label to yourself and think of everything you do in these terms

Catastrophising – viewing events as a total catastrophe, magnify things out of all proportion

Demands – I should, they must, I need to, I have to – inflexible rules that we have to live by. Any deviation can cause anger and frustration. You become a harsh critic of yourself

We set ourselves impossible standards which we cannot meet

INTERPERSONAL THERAPY

- Focussed
- 12-16 sessions

- Structured and time limited

- Problem areas

  - Interpersonal disputes (arguments/ disappointments or disagreements)

  - Role transitions (changes of circumstances)

  - Grief (not getting over the death of a loved one)

  - Interpersonal deficits/sensitivity (ability to establish and maintain supportive relationships)

PSYCHODYNAMIC PSYCHOTHERAPY

- Based on conversations with therapist and the relationship that develops between the therapist and the patient

- Looks at previous relationships, experiences and conflicts and their impact on current functioning. Difficult experiences particularly those in early life give rise to psychological conflicts

- These conflicts if unresolved, lead to problems in family and personal relationships Representations of these relationships may develop during therapy and provide a route towards understanding and change

- Occurs weekly, sessions last 50 minutes

- Can last 3 months to 2 years

- Can be individual or group

- Early sessions concentrate on formulation of problems- identify conflicts and deficits in personal development

- Transference - unconsciously determined responses that you have toward your therapist which gives an idea of responses learned from your past

- Resistance - process of the patient attempting to maintain his or her status quo. Because of your life experiences and the other things you make up your personality. If you try to change this you will encounter resistance. This is normal part of treatment

COGNITIVE ANALYTIC THERAPY (CAT)

- It is concerned with repeating and self destructive patterns in people's relationships

- The first 3-4 sessions enable the therapist to get a full picture of the difficulties and their life, at the end of which the therapist will put something down on paper so that you can agree on the main issues that need to be addressed. These will become the focus of therapy

- Also you will both be drawing some diagrams of the sequences of your behaviour and feelings. At the end of your therapy your therapist will ask you to reflect on the experience and changes you have made and you will both exchange goodbye letters

- Meet weekly for 16 sessions, each session lasts 50 minutes

- Therapy involves some questionnaires, keeping a diary and monitoring what causes you difficulties or what works well

- You will be asked to bring these observations to the sessions and look at what you have learnt about yourself with the therapist

SUITABILITY FOR PSYCHOTHERAPY (GENERAL)

- Psychologically minded and able to think in a psychological way and reflect on experiences and feelings

- Motivation to change problems and able to collaborate with therapist

- Can form and maintain a therapeutic relationship with therapist

- Able to identify focussed goals and work towards them

- Can deal with change and frustration

- "Ego strength"

CHAPTER 16   MISCELLANEOUS

PSYCHOSEXUAL HISTORY

This is a sensitive subject which needs patient's consent. Some candidsates find this difficult, but don't forget you have a task in hand and you need to perform professionally. Ask the patient if they would like a chaperone present.

- Current relationship

- Sexual orientation

- Quality/duration of relationship – Do you feel satisfied with your sex life?

- Sexual behaviour – libido/frequency/preference/practices

- Satisfaction with intercourse where appropriate

- Contraception

- Aggression/violence

- Previous relationships

- Overall number/ longest lasting/ sexual orientation

- Reasons for breakdown

- Quality

- If heterosexual ask about homosexual fantasies

- Fantasies/fidelity

- Sexual history

    - Age of first sexual activity

    - Age of first sexual intercourse

- Atittudes

- Ask about developing secondary sexual characteristics- pubic hair, breasts etc

- History of sexual problems- Premature ejaculation, erectile dysfunction

- History of sexual abuse

CAPACITY ASSESSMENT

- Brief history relating to current difficulties

- Assess mental state and determine if impacting on patients ability to consent

- Understand the information

- Assess benefits, risks and alternatives

- Weigh up the information to arrive at a decision (and understand alternatives)

- Retain and communicate the decision

- Assess cognition
    - Rule out delirium and dementia, learning disability
    - Other factors- active symptoms of mental illness, lack of emotional maturity

INSOMNIA

- Up to 15% of general population

- F>M, more in elderly

- Difficulty falling asleep (taking >30mins or more) or maintaining sleep

- Recurrent waking during the night

- Waking early in the morning and not being able to return to sleep

- Feeling unrefreshed after a nights sleep of an apparently adequate duration

- Occurs 3 times a week for at least one month

- Preoccupation with sleeplessness

- Causes distress and affects functioning

- Sleep hygiene

    - Consistent time, don't have a clock near the bed so you cant check

    - Don't lie awake thinking about your problems

    - Bedroom for sleep and sexual acitivity only (not work)

    - Comfortable bed, room at right temperature, no disturbance by a sleeping partner

    - During the day avoid caffeine, alcohol, smoking, avoid naps

    - Avoid large meals before bed

    - Healthy lifestyle (diet, exercise, regular exposure to sunlight)

- Treatment

    - CBT

    - Relaxation techniques

- Medications

    - Intermittent dosing (2-4 times a week)

    - Lowest dose

    - Short term (2 weeks)

    - Discontinue gradually (as can get withdrawal- tremor, involuntary movements, tinnitus, nausea, vomiting, insomnia and disturbing dreams)

    - Don't work for very long

    - Develops tolerance

    - some people become addicted to them

    - Side effects with BZD - hangover effects, affect alertness, memory, mood

    - Side effects non BZD - paradoxical stimulation, GI disturbance, metallic taste with zopiclone, dependency

TESTAMENTARY CAPACITY

- What is a will

- What is the consequence of leaving a will

- How would you go about making a will

- What if you want to change will

- What are your assets- e.g.: property and its worth

- Who you are giving it to

- Rule out mental health problems- psychosis, affective symptoms, cognitive impairment

SEROTONIN SYNDROME

- History of previous antidepressants, dose and duration of treatment
- Current antidepressant, dose, duration and treatment
- Length of washout
- Previous reactions
- 1% risk for SSRIs
- Symptoms
    - Fever
    - Tachycardia
    - Tremor
    - Myoclonus
    - Hyperreflexia
    - Confusion
    - Diarrhea
    - Agitation and restlessness
- Investigations- FBC, U & Es, LFTs, glucose, ABG, calcium, magnesium, phosphate. anion gap, creatinine kinase, drug toxicology, urine (myoglobinuria)
- Treatment

- Stop serotonergic drugs

- Discuss with medical colleagues and transfer

- Close monitoring of vital signs

- Hydration

- Cooling blankets for hyperthermia

- Anticonvulsants for seizures

- Clonazepam for myoclonus

- Nifedipine for hypertension

- Intramuscular Chlorpromazine as antipyretic and sedative

NEUROLEPTIC MALIGNANT SYNDROME

- Rare reaction to antipsychotic medication- secondary to dopamine activity in different parts of the brain

- 0.07%- 2%

- Risk factors include rapid antipsychotic escalation, withdrawal of antiparkinsonian medication, previous NMS, dehydration

- Fever, rigidity, confusion, labile blood pressure, raised WCC and ESR, dysphagia, incontinence, mutism

- Unpredictable, potentially life threatening (12-18% mortality)

- Investigations- FBC, U & Es, LFTs, glucose, ABG, calcium, magnesium, phosphate, anion gap, CK, drug toxicology, CXR, ECG, blood cultures, CT head and LP, EEG, urine (myoglobin)

- Treatment

    - Transfer to medical setting

    - Discontinue neuroleptics

    - Benzodiazepine to control agitation

    - Bromocriptine/Sodium dantrolene to relax muscles

    - IV fluids, antipyretic, cooling blankets for hyperthermia

    - Course: 7-10 days

    - Once settled- drug holiday of 2 weeks

    - Rechallenge with atypical antipsychotic with careful monitoring of pluse, blood pressure, temperature, consciousness levels and creatinine kinase

    - ECT can be used

- 1 in 6 will have recurrence

DYSTONIA/OCULOGYRIC CRISIS

- Acute reaction after exposure to antipsychotic medication with sustained painful muscular spasms, producing twisted abnormal movements

- Aetiology unknown

- More likely in young males with previous history or family history of dystonia

- Occurs in approximately 10%

- Stop suspected agent

- IM/IV procyclidine 5 mgs in emergency

- May feel dry mouth, blurred vision, dizziness, possible sedation due to anticholinergic

- May need regular anticholinergic prophylaxis for patients with history of drug induced dystonia

CHAPTER 17

INCIDENCE, PREVALENCE, PROGNOSIS

DEMENTIA

- After the age of 65- the risk doubles every 5 years

- Over the age of 65, six in 100

- Over the age of 80, 1 in 5

- Live for 5-10 years

- Antidementia drugs: 50-60% efficacy

- AD: 3% of people over the age of 65, over the age of 85 it rises to 10-15/100

- CJD: affects 1 in a million people in UK

- CJD: progressive death within 1-2 years

- Huntington's- death in 10-15 years

- Lewy Body

    - 4% of all dementias

    - 50% sensitivity reactions with neuroleptics

- Parkinson's dementia- 15-20% of all Parkinson's disease

- Olanzapine increases the risk of stroke by 3-4 times and the incidence of stroke is 1-2 in 100

DELIRIUM

- 30% of all elderly medical patients

- Mortality- 6-18%

- Up to a third is preventable cause

- Patients with dementia are 5 times more likely to develop delirium

- Two third of cases can be missed

ALCOHOL

Delirium Tremens:

- 5% of withdrawal

- Epileptic seizures- one third

- 80% end within 72 hours

- Mortality: 5- 15%

CANNABIS

- 1 in 10 develop unpleasant experiences

- Risk of Schizophrenia doubles

- Significant high risk of depression (teenage years if smoked)

- 3 out of 4 experience craving

- Detected in urine up to 56 days after last consumption

SCHIZOPHRENIA

- Affects 1 in 100

- Depression:

    - Before treatment, around half will develop

    - 1 in 7 with continuing symptoms will have depression

- Causes:

    - 1in 10 with Schizophrenia have a parent with illness

    - Genes account for half of the risk of Schizophrenia

    - Cannabis doubles the risk

- Outlook:

    - 1 in 5 get better within 5 years of their first episode

    - 3 in 5 get better but still have some symptoms

    - 1 in 5 will continue to have troublesome symptoms

- 4 in 5 get better with meds

- Take medications for 1-2 years

- If medications stopped, risk of relapse doubles

- 80-90% will have problems with sex life

- Typical medications- 45% problems with sex life

- TD- 1in 20 people every year who take typical

- Weight gain by atypical- first 12 months

- Clozapine:

    - Neutropenia- 1%

    - Agranulocytosis- 3%

    - 6 out of 10 will benefit

DEPRESSION

- Symptoms for 2 weeks

- About 1 in 10 who suffer from serious depression will also have periods when they will be elated

- 4 out of 5 people with depression will get completely better without any help- takes 4-6 months

- Antidepressants:

    - 50-65% work

    - First episode- take for 6 months after you get better

    - 2 episodes- 2 years

    - 2 episodes- consider longer

- ECT: 8 out of 10 people respond

- Risk of death in ECT is one in 50,000

- 1-5/10 show people have memory problems

- SSRIs in pregnancy- increased chance of preterm delivery & reduced birth weight
- Up to a third have withdrawal symptoms

ANTIDEPRESSANTS & SEXUAL DYSFUNCTION

- 40-50% of people with depression report diminished libido and problems with sexual arousal
- Tricyclics- 30%
- MAOIs- 40%
- SSRIs- 60-70%
- Venlafaxine- 70%
- Mirtazapine-25%
- Spontaneous remission- 10%
- Drug holidays, delayed dosing, reduce drugs, discontinue, adjunctive antidote drugs
- Sildenafil improves erectile function in men taking SSRIs
- Bupropion helps as well

TREATMENT RESISTANT DEPRESSION

- 10-20% of major depression

- 50% of previous non responders become responders to second antidepressant

MANIA

- Symptoms for 1 week
- Lithium:
    - Check Lithium and U & Es every 3 months
    - TFTs every 6 months
    - Takes 3 months to act
    - Weight gain- 25% gain more than 4.5 kg
    - Kidney changes: 10-20%
    - Renal failure – 1%
    - Hypothyroidism: 5-35%
    - Ebstein's anomaly: 8 fold RR
    - Congenital anomalies: 4-12%
    - Reduces relapse rate by 30-40%
    - 10-50% of Lithium is secreted in breast milk
- How long:
    - At least 2 years after an episode of Bipolar
    - For up to 5 years if there are risk factors
        - Frequent relapses

- Psychotic episodes

- Substance misuse

- Continuing social stress

BIPOLAR AFFECTIVE DISORDER

- 1 in 100 affected

- Ebstein's anomaly with Lithium: risk raised from 1 in 20000 to 1 in 2000

- Foetal heart defects with Lithium- risk raised from 8 in 1000 to 60 in 1000

- Congenital anomalies: 4-12%

- Risk of foetal malformations for pregnant women without mental disorder: 2-4/100

- 1st episode: Take medications at least 2 years

- Risk factors such as frequent relapses, psychotic episodes, substance misuse, continuing social stress- 5 years

- Pregnancy: Check Lithium every 4 weeks, then weekly from the 36th week, and less than 24 hours after childbirth

PANIC DISORDER

- 4 fold increase in suicide attempts in patients with panic disorder and depression

OCD

- OCD- For most days at least 2 weeks

- Prevalence is 2 % (twice as common as schizophrenia or bipolar disorder). One out of every 50 individuals suffer from OCD

- Stressful life events bring it on in about one out of three cases

- More than half of the clients do respond to the treatment.

- 2/3rd get depression, 1/3rd get major depression

- Medications for 12 weeks

  3 out of 4 individuals improve with exposure and response prevention

- 6 out of 10 individuals suffering from OCD improve with medication. On average symptoms reduce by half. About 1 in 2 who stops medications will get symptoms again in the months of stopping.

- Half of them may relapse if they stop medication.

- Continue medications at least for one year

ACUTE STRESS REACTION

- Onset within min/hours

- Gets better within 48 hours

- Minimal after about 3 days

PTSD

- 1 in 3 will continue to have symptoms after a traumatic event

- Trauma focussed CBT- 1st option

- Continue antidepressants for 12 months and then gradually reduce it over a period of 4 weeks

- 50% continue to have a long term problems

ADJUSTMENT DISORDER (including grief reaction)

- Onset within 1 month

- Duration of symptoms does not exceed 6 months

- Brief depressive reaction- not exceeding 1 month

- Prolonged depressive reaction- not exceeding 2 years

SOMATIZATION DISORDER

- At least 2 years of multiple and variable physical symptoms

HYPOCHONDRIASIS

- 0.8-5% incidence in medical inpatient settings

- Condition persists at 4-5 year follow up

EATING DISORDER

- Prevalence of anorexia nervosa in 15-year-old girls is 1 out of 150 teenage girls will suffer from anorexia nervosa. In boys, it is 1 boy out of 1000.

- More than half recover

- Full recovery can occur after 20 years

- 1 in 5 can die

- About 4 out of every 100 woman suffer from Bulimia nervosa

- On average duration of illness is 6 years for AN

POSTNATAL DEPRESSION

- 1 in 10 mothers become depressed after having babies.

- Baby blues- half of mothers, on the 3rd of 4th day

- Chances of getting Postnatal Depression again- without a h/o depression- 10-15%; After one episode of PND- 20—40%

- Tobacco In pregnancy- 3 times more likely to have a baby with low birth weight. Stillbirths and neonatal deaths are 30% higher

PUERPERAL PSYCHOSIS

- 1 in 500 mothers

- Previous H/O BPAD- 50% chance

- 1 in 5 will have puerperal psychosis again

- Half of mothers will not have any problems

PERSONALITY DISORDER

- 40-70% in psychiatric ward

- 30-40% of patients in community

- 10-30% of patients who see GP

MENTAL RETARDATION

- Mild: 50-69

- Moderate: 35-49

- Severe: 20-34

- Profound: under 20

AUTISM

- Childhood: less than 3 years, in all 3 areas

- Atypical: after age 3, 1-2 of the three areas

- Rett's syndrome: 7-24 months

- Childhood disintegrative disorder: Normal dev for 2-3 years, followed by loss of acquired skills (motor, language and social) between 3-4 years

NEUROLEPTIC MALIGNANT SYNDROME (NMS)

- Roughly 1 in 1000 people who are on any antipsychotic will develop NMS.
- Less than 1 in 100 with conventional antipsychotics
- Up to 90% do get better while 10-15% die from NMS
- Treat with Quetiapine or Clozapine
- 1 in 6 recurrence

ADHD

- Onset before the age of 6 years
- Co morbidity:
    - Specific learning disorders- 60%
    - Conduct Disorder and Oppositional Defiant Disorder- 40%
- Measure height every 6 months and weight every 3 months
- Outcome:
    - 20%- dissocial personality traits
    - 15-20%- substance misuse
    - 20-30%- full ADHD syndrome persists in to adulthood

- 60%- One or more symptoms persist in to adulthood

CONDUCT DISORDER

- Symptoms for 6 months

PHYSICAL HEALTH

- 1 in 10 children suffer from physical symptoms for which no medical cause can be found.
- School phobia/ refusal- one third develop psychiatric disorders in adulthood

INSOMNIA

- At any one time affects one third of adults
- 10- 15% will have problems for months- years

SCHOOL REFUSAL

- One third develop psychiatric disorders
- One in four primary school pupils/ one in 10 secondary school children are being bullied

REFERENCES

1.   Royal College of Psychiatrists. The MRCPsych Clinical Assessment of Skills and Competencies (CASC): Candidate Guide, London: Royal College of Psychiatrists, 2009

2.   Royal College of Psychiatrists. MRCPsych CASC Pass Mark London: Royal College of Psychiatrists, 2009

3.   Wing JK, Cooper JE, Sartorius N, The measurement and classification of psychiatric symptoms: An instruction manual for the Present State Examination and CATEGO programme Cambridge: Cambridge University Press, 1974

4.   Cayton H, Graham N, Warner J. Dementia, Alzheimer's and other dementias. London: Class Publishing, 2004

5.   Hill L, Briscoe M. Drug treatment of Alzheimer's disease. London: Royal College of Psychiatrists, 2009

6.   Hill L, Briscoe M. Memory problems and dementia. London: Royal College of Psychiatrists, 2009

7.   Folstein MF, Folstein SE, McHugh PR. "Mini-mental state" A practical menthod for grading the cognitive state of patients for the clinician. Journal of Psychiatric Research 1975;12(3): 189-198

8.   Royal College of Psychiatrists Dementia:Key Facts. London: Royal College of Psychiatrists, 2009

9.   Royal College of Psychiatrists Delirium. London: Royal College of Psychiatrists, 2009

10.   World Health Organization The ICD-10 Classification of Mental and Behavioural Disorders: Clinical Descriptions and Diagnositc Guidelines (CDDG) Geneva: WHO, 1992

11.   National Institute for Health and Clinical Excellence, Social care institute for excellence. Dementia: Supporting peoplen with dementia and their carers in health and social care, NICE, 2006

12.   The Birmingham Course. Clinical Assessment of Skills and Competencies (CASC) Course. Birmingham: The Birmingham Course, 2009

13.   Cambridge Course. CASC course. Cambridge: Cambridge Course, 2009

14.   Christmas D. OSCEs Stations. www.trickcylists.co.uk, 2007

15.   Royal College of Psychiatrists. Alcohol:Our Favourite Drug. London: Royal College of Psychiatrists, 2008

16.   Royal College of Psychiatrists. Alcohol and Depression. London: Royal College of Psychiatrists, 2008

17.   Royal College of Psychiatrists. Alcohol and Older People. London: Royal College of Psychiatrists, 2008

18. Royal College of Psychiatrists. Cannabis and mental health. London: Royal College of Psychiatrists, 2008

19. National Institute for Health and Clinical Excellence. Drug misuse: opiod detoxification, NICE, 2007

20. National Institute for Health and Clinical Excellence. Drug misuse: psychosocial interventions, NICE, 2007

21. Royal College of Psychiatrists. Schizophrenia – Key Facts. London: Royal College of Psychiatrists, 2008

22. Royal College of Psychiatrists. Depot medication. London: Royal College of Psychiatrists, 2007

23. Royal College of Psychiatrists. Schizophrenia. London: Royal College of Psychiatrists, 2004

24. Royal College of Psychiatrists. Cannabis and mental health. London: Royal College of Psychiatrists, 2008

25. Royal College of Psychiatrists. Severe mental illness (psychosis). London: Royal College of Psychiatrists, 2005

26. National Institute for Health and Clinical Excellence. Guidances on the use of newer(atypical) antipsychotic drugs for the treatment of schizophrenia. NICE technology appraisal guidance 43, NICE, 2002

27. National Institute for Health and Clinical Excellence. Schizophrenia:core interventions in the treatment and management of schizophrenia in primary and secondary care. NICE clinical guideline 1, NICE, 2002

28. National Institute for Health and Clinical Excellence. Schizophrenia:core interventions in the treatment and management of schizophrenia in primary and secondary care. NICE clinical guideline 1 (update), NICE, March 2009.

29. Royal College of Psychiatrists. Depression – Key Facts. London: Royal College of Psychiatrists, 2008

30. Royal College of Psychiatrists. Depression. London: Royal College of Psychiatrists, 2008

31. Royal College of Psychiatrists. Alcohol and Depression. London: Royal College of Psychiatrists, 2008

32. Royal College of Psychiatrists. Depression in Older Adults. London: Royal College of Psychiatrists, 2008

33. Royal College of Psychiatrists. Men and Depression. London: Royal College of Psychiatrists, 2008

34. Royal College of Psychiatrists. Patients and antidepressants. London: Royal College of Psychiatrists, 2008

35. Royal College of Psychiatrists. Self-Harm. London: Royal College of Psychiatrists, 2007

36. National Institute for Health and Clinical Excellence. Depression: the treatment and management of depression in adults, NICE, 2004

37. National Institute for Health and Clinical Excellence. Depression: the treatment and management of depression in adults (update), NICE, October 2009

38. National Institute for Health and Clinical Excellence. Depression in children and young people: the ideintification and managment in primary, community and secondary care, NICE, 2005

39. National Institute for Health and Clinical Excellence. Depression: the treatment and management of depression in adults, NICE, 2004

40. Royal College of Psychiatrists. Eating Disorders. London: Royal College of Psychiatrists, 2008

41. Royal College of Psychiatrists. Eating disorders: Key Facts. London: Royal College of Psychiatrists, 2009

42. National Institute for Health and Clinical Excellence. Eating disorders: Core interventions in the treatment and management of anorexia nervosa, bulimia nervosa and related eaing disorders, NICE, 2004

43. Stein G. (ed.) Seminars in General Adult Psychiatry. London: Royal College of Psychiatrists, 2007

44. Royal College of Psychiatrists. Anxiety & Phobias. London: Royal College of Psychiatrists, 2006

45. Royal College of Psychiatrists. Coping with trauma: how to cope after a traumatic event. London: Royal College of Psychiatrists, 2006

46. Royal College of Psychiatrists. Shyness and Social Phobia. London: Royal College of Psychiatrists, 2008

47. Royal College of Psychiatrists. Anxiety, Panic & Phobias: Key Facts. London: Royal College of Psychiatrists, 2008

48. Royal College of Psychiatrists. Benzodiazepines. London: Royal College of Psychiatrists, 2009

49. National Institute for Health and Clinical Excellence. Anxiety: management of anxiety (panic disorder, with or without agoraphobia, and generalised anxiety disorder) in asults in primary,secondary and community care, NICE, 2004

50. National Institute for Health and Clinical Excellence. Obsessive compulsive disorder (OCD) and body dysmorphic disorder (BDD) : core interventions in the treatment of obsessive compulsive disorder and body dysmorphic disorder, NICE, 2005

51. National Institute for Health and Clinical Excellence. Post-traumatic stress disorder (PTSD): the manangement oif PTSD in adults and children in primary and seconary care, NICE, 2005

52. Royal College of Psychiatrists. Anxiety & Phobias. London: Royal College of Psychiatrists, 2006

53. Royal College of Psychiatrists. Coping with trauma: how to cope after a traumatic event. London: Royal College of Psychiatrists, 2006

54. Royal College of Psychiatrists. Shyness and Social Phobia. London: Royal College of Psychiatrists, 2008

55. Royal College of Psychiatrists. Anxiety, Panic & Phobias: Key Facts. London: Royal College of Psychiatrists, 2008

56. Royal College of Psychiatrists. Benzodiazepines. London: Royal College of Psychiatrists, 2009

57. National Institute for Health and Clinical Excellence. Anxiety: management of anxiety (panic disorder, with or without agoraphobia, and generalised anxiety disorder) in asults in primary,secondary and community care, NICE, 2004

58. National Institute for Health and Clinical Excellence. Obsessive compulsive disorder (OCD) and body dysmorphic disorder (BDD) : core interventions in the treatment of obsessive compulsive disorder and body dysmorphic disorder, NICE, 2005

59. National Institute for Health and Clinical Excellence. Post-traumatic stress disorder (PTSD): the manangement oif PTSD in adults and children in primary and seconary care, NICE, 2005

60. Royal College of Psychiatrists. Postnatal Depression. London: Royal College of Psychiatrists, 2007

61. Royal College of Psychiatrists. PND: Key Facts. London: Royal College of Psychiatrists, 2008

62. Royal College of Psychiatrists. Mental illness after childbirth. London: Royal College of Psychiatrists, 2008

63. National Institute for Health and Clinical Excellence. Antenatal and postnatal mental health: clinical management and service guidance, NICE, 2007

64. Royal College of Psychiatrists, Stoddart J. Personality Disorders. London: Royal College of Psychiatrists, 2007

65. Royal College of Psychiatrists. Personality Disorders: Key Facts. London: Royal College of Psychiatrists, 2009

66. Webster CD, Douglas KS,Eaves D, et al HCR-20:Assessing Risk for Violence (Version 2).Toronto: Psychological Assessment Resources, 1997

67. Royal College of Psychiatrists. Sleeping well. London: Royal College of Psychiatrists, 2009.

68. Lindsay KW, Bone I, Callander R. Neurology and neurosurgery illustrated London: Churchill Livingstone, 1997

69. Cohen RM. Patient management problems for the MRCPsych. Trowbridge: Cromwell Press, 2004.

70. Zoha M, Lowe J, Wise J, Gosall G, Kaligotla S. MRCPsych OSCEs for Part 1. Knutsford:Pastest, 2004.

71. Michael A. (ed.) OSCEs in psychiatry: prepare for the the new MRCPsych. Edinburgh: Churchill Livinstone, 2004.

72. Rao R.(ed.) OSCEs in Psychiatry. London: Gaskell, 2005

73. Olumoroti OJ, Kassim AA. Patient management problems in psychiatry. Edinburgh: Elsevier Churchill Livingstone, 2005.

74. Murthy SPM. Get through MRCPsych part 1: preparation for the OSCEs. London: Royal Society of Medicine Press, 2004

75. Folstein MF, Folstein SE, McHugh PR. Mini-mental state: a practical method for grading the cognitive state of patients for the clinician. Journal of Psychiatric Research, 1975;12:189-198

76. Dubois B, Slachevsky A, Litvan I, Pillon B. The FAB: a frontal assessment battery at bedside. Neurology 2000;55(11):1621-1626.

77. Semple D, Smyth R, Burns J, Darjee R, McIntosh A. Oxford handbook of psychiatry, New York: Oxford University Press, 2005.

INDEX

1

11th nerve (Accessory), 222
12th nerve (Hypoglossal), 223
1st nerve (olfactory), 221

2

2nd nerve (optic), 221

3

3rd, 4th and 6th nerve, 222

5

5th nerve (Trigeminal), 222

7

7th nerve (Facial), 222

8

8th nerve (Vestibulocochlear), 222

9

9th and 10th nerve (Glossopharyngeal and vagus), 222

A

AA, 66
ABDOMINAL EXAMINATION, 219
Abnormal grief, 141
ABOUT THE AUTHORS, 9

Abstract reasoning, 53
Abstract similarities, 53
Acamprosate, 66
ACUTE AND TRANSIENT PSYCHOTIC DISORDERS, 74
ACUTE STRESS REACTION, 134
ADHD – HISTORY, 202
ADJUSTMENT DISORDERS, 139
agnosia, 32
AGORAPHOBIA, 118
Al anon, 66
ALCOHOL AND DRUGS, 63
ALCOHOL MISUSE PHYSICAL EXAMINATION, 227
ALCOHOL WITHDRAWAL-MANAGEMENT, 65
Alternate sequence, 53
ALZHEIMERS DEMENTIA, 27
Anankastic, 176
Animal assisted therapy, 41
ANOREXIA NERVOSA, 149
ANOREXIA NERVOSA-EXPLANATION, 151
ANTIDEMENTIA DRUGS, 37
Antidepressants, 101
ANTIDEPRESSANTS AND BREASTFEEDING, 109
ANTIDEPRESSANTS AND PREGNANCY, 109
ANTIDEPRESSANTS IN PREGNANCY, 168
ANTIPSYCHOTICS IN DEMENTIA, 41

ANTIPSYCHOTICS IN PREGNANCY, 171
Antisocial (Dissocial), 175
ANXIETY AND SOMATOFORM DISORDERS, 117
ANXIETY SYMPTOMS, 117
Anxious (avoidant), 175
ANXIOUS (AVOIDANT) PERSONALITY DISORDER, 120
Arbitrary inference, 238
Aricept, 37
Aromatherapy, 41
Asomatognosia, 61
ASPERGER'S SYNDROME, 196
ASSAULT ON STAFF, 188
Astereognosia, 61
ATOMOXETINE, 208
Auditory hallucinations, 19
Augmentation with antipsychotics, 108
Augmentation with lithium, 108
Aura, 49
Autism Diagnostic Observational Schedule (ADOS), 199
Automatisms, 50
Autonomic arousal symptoms, 117

B

BEHAVIOURAL SYMPTOMS IN DEMENTIA, 45
BENZODIAZEPINES IN PREGNANCY, 170

270

BIPOLAR AFFECTIVE DISORDER, 88
BIPOLAR AFFECTIVE DISORDER (BPAD)- EXPLANATION, 88
BODY DYSMORPHIC DISORDER- HISTORY, 133
BREAKING BAD NEWS, 62
BULIMIA NERVOSA, 158
BULIMIA NERVOSA- EXPLANANTION, 158
BULLYING, 213

C

CAGE questionnaire, 64
CANNABIS AND MENTAL HEALTH, 70
CAPACITY ASSESSMENT, 244
CARBAMAZAPINE IN PREGNANCY, 170
CARBAMAZEPINE, 92
CARDIOVASCULAR SYSTEM EXAMINATION, 217
Catastrophising, 238
CHILDHOOD AUTISM, 196
CHILDHOOD DISINTEGRATIVE DISORDER, 202
CHILDHOOD DISORDERS, 195
CHILDREN WITH PSYCHOSIS, 195
CHRONIC FATIGUE SYNDROME, 146
Clozapine, 78
COCAINE AND PREGNANCY, 69
COGNITIVE ANALYTIC THERAPY (CAT), 241

COGNITIVE BEHAVIOURAL THERAPY, 234
COGNITIVE DISTORTIONS, 236
Cognitive Estimates, 53
Cognitive symptoms, 31
Combinations of antidepressants, 108
Communication difficulties, 197
Comprehension, 32
Compulsions, 130
CONDUCT DISORDER, 210
Constructional dyspraxia, 61
CONTENTS, 11
CRANIAL NERVE EXAMINATION, 221
Cyclothymia, 89, 99

D

Déjà vu, 49
DELIRIUM, 29
Delusion, 16
DELUSIONAL DISORDER, 73
Delusional jealousy, 18
DELUSIONAL JEALOUSY- HISTORY AND MANAGEMENT, 83
Delusional mood, 16
Delusional perception, 16
DELUSIONS, 15
Delusions of control, 16
Delusions of guilt, 17
Delusions of persecution, 17
DEMENTIA AND WANDERING – ASSESS RISKS, 46

DEMENTIA- HISTORY TAKING, 31
DEMENTIA IN HUNTINGTON'S DISEASE, 28
DEMENTIA IN PICK'S DISEASE, 28
DEPENDENCE SYNDROME, 63
Dependent, 176
DEPRESSION AND PHYSICAL HEALTH PROBLEMS, 110
DEPRESSION AND SEXUAL SIDE EFFECTS OF MEDICATIONS, 105
DEPRESSION- EXPLANATION, 99
DEPRESSIVE EPISODE, 97
DEVELOPMENTAL HISTORY, 195
Dichotomous thinking, 238
DIFFERENTIAL DIAGNOSIS FOR ANXIETY SYMPTOMS, 120
DISSOCIATIVE AMNESIA, 143
DISSOCIATIVE ANAESTHESIA AND SENSORY LOSS, 144
DISSOCIATIVE DISORDERS, 142
DISSOCIATIVE FUGUE, 143
DISSOCIATIVE MOTOR DISORDER, 144
Disulfiram, 66
Donepezil, 37
DSH- RISK ASSESSMENT, 23

DUAL ROLE TRANSVESTITISM, 183
Dysgraphaesthesia, 61
dysgraphia, 32
Dyslexia, 32
Dyspraxia, 32
Dysthymia, 99
DYSTONIA/OCULOGYRIC CRISIS, 249

E

EATING DISORDER- HISTORY, 149
EATING DISORDERS, 149
EATING DISORDERS EXAMINATION, 228
Ebixa, 39
Ebstein's anomaly, 91
ECT- EXPLANATION, 111
Elderly mentally ill nursing, 49
Elderly mentally ill residential, 49
ELICIT HISTORY OF ALCOHOL DEPENDENCE, 63
EMDR (Eye movement desentisation and reprocessing), 139
Emotional reasoning, 238
Emotionally unstable - borderline type, 177
Emotionally unstable- impulsive type, 177
EROTOMANIA, 84
Exelon, 37
EXHIBITIONISM, 183
EXPLAIN ALZHEIMER'S DISEASE, 43
EXPLAIN FRONTO- TEMPORAL DEMENTIA, 45

EXPLAIN LEWY BODY DEMENTIA, 45
EXPLAIN TEMPORAL LOBE EPILEPSY, 49
EXPLAIN VASCULAR DEMENTIA, 44
Exposure and response prevention, 131
EXTRAPYRAMIDAL SIDE EFFECTS EXAMINATION, 221

F

FAS test, 52
Finger agnosia, 61
FIRST RANK SYMPTOMS, 20
FITNESS TO PLEAD, 186
FOR CARERS, 95, 105, 132
Forced thinking, 49
FORENSIC PSYCHIATRY, 185
frontal lobe features, 28
FRONTAL LOBE SYNDROME, 30
Frontal release signs, 53

G

Galantamine, 37
GENDER IDENTITY DISORDER OF CHILDHOOD, 183
GENERALISED ANXIETY DISORDER, 119
GENERALISED ANXIETY DISORDER- TREATMENT, 126
Ginko Biloba, 40
Go-no-go test, 53

GRADED EXPOSURE WITH RELAXATION, 125
Grandiose delusions, 17
Grasp reflex, 54
GRIEF REACTION, 140

H

HALLUCINATIONS, 18
HCR 20, 192
HEBEPHRENIC SCHIZOPHRENIA, 72
High dose monotherapy, 107
Histrionic, 176
Hyperthyroidism, 224
HYPOCHONDRIACHAL DISORDER, 145
Hypochondriasis, 18
HYPOMANIA, 87
Hypothyroidism, 224

I

Ideas of reference, 15
INCIDENCE, PREVALENCE, PROGNOSIS, 251
INDECENT EXPOSURE, 185
INSIGHT, 22
INSOMNIA, 245
INTERPERSONAL THERAPY, 239

J

Jamais vu, 49

K

KORSAKOFF'S ASSESSMENT, 54

KORSAKOFF'S SYNDROME, 28

L

Labelling, 238
LAMOTRIGINE, 93
LAMOTRIGINE IN PREGNANCY, 171
Language difficulties, 32
LEARNING DISABILITY, 193
LEWY BODY DEMENTIA, 30
LEWY BODY DEMENTIA MANAGEMENT, 42
LITHIUM, 90
LITHIUM AND PREGNANCY, 91
LITHIUM IN PREGNANCY, 169
LOWER LIMB NEUROLOGICAL EXAMINATION, 226
Luria's test, 53

M

Made acts, 21
Made affect, 21
Made will, 21
Magnification, 237
MANAGEMENT OF CHALLENGING BEHAVIOUR IN DEMENTIA, 55
MANAGEMENT OF DEMENTIA- DISCUSS WITH CONSULTANT, 34
MANIA WITH PSYCHOTIC SYMPTOMS, 88

MANIA WITHOUT PSYCHOTIC SYMPTOMS, 87
Massage, 41
Memantine, 39
memory (short and long term), 31
MENTAL RETARDATION, 193
METHYLPHENIDATE, 207
MILD COGNITIVE DISORDER, 29
Mild depression, 100
Mild depressive episode, 97
MILD MENTAL RETARDATION, 193
MINI MENTAL STATE EXAMINATION (MMSE), 60
Minimization, 237
Moderate depressive episode, 97
MODERATE MENTAL RETARDATION, 193
MODERATE TO SEVERE DEPRESSION IN CHILDREN, 108
MOOD DISORDERS, 87
MOTIVATIONAL INTERVIEWING, 67
Motor sequencing, 53
Multisensory stimulation, 41
Mummification, 141

N

Narcissistic, 176
Negative symptoms, 71
NEURASTHENIA, 146
NEUROLEPTIC MALIGNANT SYNDROME, 248

NICE GUIDELINES- TREATMENT OF BPAD, 96
Nihilistic delusions, 18
NON EPILEPTIC SEIZURES (NES)- INFORMATION TO CARER, 57
NON-EPILEPTIC SEIZURES- HISTORY, 56
Normal grief, 140

O

Obsessions, 129
OBSESSIVE COMPULSIVE DISORDER, 126
OCD- EXPLANATION, 129
OCD- HISTORY, 126
OCD TREATMENT, 128
Olfactory hallucinations, 20
OPIATE USE AND PREGNANCY, 68
OPIATE WITHDRAWAL EXAMINATION, 230
OPPOSITIONAL DEFIANT DISORDER, 212
ORGANIC AMNESIC SYNDROME, 28
ORGANIC MENTAL DISORDERS, 27
ORGANIC PERSONALITY DISORDER, 30
overdose following sexual assault or rape, 25
Overgeneralization, 237
Oxazepam, 65

P

PAEDOPHILIA, 184

PAEDOPHILIA- HISTORY, 187
Palmomental reflex, 54
PANIC ATTACKS- EXPLANATION, 124
PANIC DISORDER, 119
PANIC DISORDER- MANAGEMENT (NICE GUIDANCE), 125
Paranoid, 174
PARANOID SCHIZOPHRENIA, 72
PARAPHRENIA, 82
PARENT TRAINING, 209
Parenteral thiamine, 65
PARIETAL LOBE TESTING, 61
PATHOLOGICAL FIRESETTING (PYROMANIA), 181
PATHOLOGICAL GAMBLING, 181
PATHOLOGICAL STEALING (KLEPTOMANIA), 182
Personalisation, 238
PERSONALITY DISORDER - EXPLANATION, 178
PERSONALITY DISORDERS, 173
PHYSICAL EXAMINATIONS, 217
Poor prognostic factors, 157, 159
POST HEAD INJURY ASSESSMENT, 50
POST HEAD INJURY ASSESSMENT OF COGNITIVE FUNCTION, 52
POST NATAL DEPRESSION & PUERPERAL

PSYCHOSIS- HISTORY, 161
POST STROKE DEPRESSION, 110
POST TRAUMATIC STRESS DISORDER, 135
POSTNATAL DEPRESSION- EXPLANATION, 163
PREFACE, 5
PREGNANCY & LD STATION, 194
PREMORBID PERSONALITY, 173
PROFOUND MENTAL RETARDATION, 194
Proverb interpretation, 53
PSEUDOCYESIS, 162
PSYCHODYNAMIC PSYCHOTHERAPY, 240
PSYCHOPATHOLOGY, 15
PSYCHOSEXUAL HISTORY, 243
PSYCHOTHERAPY, 233
PTSD- EXPLANATION, 137
PUERPERAL DISORDERS, 161
PUERPERAL PSYCHOSIS- EXPLANATION, 166

R

Rapid cycling, 89
Recurrent depressive disorder, 98
REFERENCES, 265
Regular laxative use, 153
Religious delusions, 17
Reminiscence therapy, 41
Reminyl, 37

RESPIRATORY SYSTEM EXAMINATION, 218
RETT'S SYNDROME, 201
RISK OF OFFENDING, 190
Rivastigmine, 37
Running commentary, 20

S

Schizoid, 174
SCHIZOPHRENIA, 71
SCHIZOPHRENIA- EXPLANATION, 74
Schizotypal, 174
SCHIZOTYPAL DISORDER, 72
SCHOOL REFUSAL, 214
Selective abstraction, 237
Selegiline, 41
SELF HELP, 68, 95, 104, 132, 154, 165
Self-help, 105, 122
SEMISODIUM VALPROATE, 92
SEPARATION ANXIETY DISORDER OF CHILDHOOD, 212
SEROTONIN SYNDROME, 247
Severe depressive episode with/ without psychotic symptoms, 98
SEVERE MENTAL RETARDATION, 194
Sheltered housing, 48
snoezelan room, 41
SNRIs, 103
SOCIAL PHOBIA, 119
SOCIAL PHOBIA- EXPLANATION, 121
Socialising difficulties, 197
Somatic passivity, 21
Somatic syndrome, 98

SOMATIZATION DISORDER, 144
SOTP, 188
Spatial orientation, 31
SSRIs, 102
ST. JOHNS WORT, 103
Starvation, 153
SUGGESTIONS FOR PRACTICE, 6
SUICIDE, 23
   INDICATORS OF SERIOUSNESS, 24
SUICIDE- HIGH RISK FACTORS, 25
SUITABILITY FOR PSYCHOTHERAPY (GENERAL), 242
Switching antidepressants, 107
SYSTEMATIC DESENSITIZATION, 125

T

Tactile hallucinations, 20

Temporal orientation, 31
TESTAMENTARY CAPACITY, 246
Therapeutic use of music and dance, 41
Third person auditory hallucinations, 20
Thought alienation, 21
Thought block, 21
Thought broadcast, 21
Thought echo, 21
Thought insertion, 21
Thought withdrawal, 21
THYROID EXAMINATION, 223
TOBACCO AND PREGNANCY, 69
TRANSSEXUALISM, 182
TREATMENT RESISTANT DEPRESSION, 106
TRICYCLICS, 103

U

Unusual behaviour, 198
UPPER LIMB NEUROLOGICAL EXAMINATION, 225

V

VALPROATE IN PREGNANCY, 170
VASCULAR DEMENTIA, 27
Verbal fluency, 52
Very sheltered housing, 48
Visual hallucinations, 20
Vitamin E, 40
Vomiting, 153

W

Wernicke's encephalopathy, 54